Biochemical Systematics and Evolution

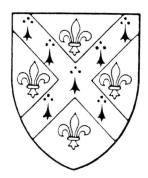

Biochemical Systematics and Evolution

ANDREW FERGUSON, B.Sc., Ph.D.

Lecturer in Zoology
The Queen's University of Belfast

Blackie

Glasgow and London

Blackie & Son Limited
Bishopbriggs
Glasgow G64 2NZ

Furnival House
14/18 High Holborn
London WC1V 6BX

First published 1980

*International Standard Book Number
0 216 90779 9*

Filmset and printed in Great Britain by
Thomson Litho Ltd, East Kilbride, Scotland

Preface

It is my sincere belief that systematics is one of the most important and indispensable, one of the most active and exciting, and one of the most rewarding branches of biological science. I know of no other subject that teaches us more about the world we live in.

(Mayr, 1969a)

Nothing in biology makes sense except in the light of evolution. (Dobzhansky, 1973)

AS THE ABOVE QUOTATIONS FROM TWO OF THE MOST EMINENT AUTHORITIES in this field indicate, a knowledge of systematics and evolution is fundamental to the study of all other areas of biology. Recent advances in biology—especially in molecular studies, electron microscopy and physiology—have lured biologists from systematic work, with the result that it has become one of the most neglected areas and, until relatively recently, was an unfavoured element in biology courses. However, the contributions of molecular biology—in particular the discovery of DNA as the carrier of genetic information—have re-vitalized systematic and evolutionary work. Thus in the field of biochemical systematics the oldest biological discipline, systematics, is combined with the newest, molecular biology.

Over the past decade there has been an enormous explosion of literature dealing with biochemical aspects of systematics and evolution. Much of this work has concerned the application of gel electrophoresis and it is to this technique that the greater part of this book is devoted. One aim is to provide the reader with the necessary background on the principles and practice of electrophoresis, to be able to understand and evaluate critically literature in this area.

With the large amount of information available it is obviously not possible to provide a complete review; such coverage would require a book larger than this for any class of vertebrates. Examples have been picked which illustrate particular concepts or applications. The reader will find that most examples concern animals, and particularly the vertebrates—in part because of the greater amount of information on this group, and also because of the author's own interests.

Little previous knowledge is assumed, other than that which would be obtained in an introductory biology course. Clearly, exposure to courses in biochemistry and genetics would facilitate quicker understanding of parts, but it is hoped that sufficient explanation of underlying concepts is provided to obviate the need for such a background. Although this book is aimed primarily at the advanced undergraduate level, the post-graduate and research worker should find parts of interest and benefit. To this end, supplementary reading is provided not only in the form of Further Reading for each chapter, but also by extensive bibliographical citations in the text. Practical details are provided, and the potentialities and limitations of electrophoretic techniques are stressed. Throughout, systematics is interpreted in the widest sense and, as well as traditional aspects, newer calls on systematic services are also considered. For example, chapter five is devoted to intra-specific systematics which covers the characterization and delimitation of populations and other groups within species. This is a fast-growing area of use of biochemical techniques, as such information is vital to ecologists, medical and fishery biologists, etc.

I would like to express my gratitude to the many people who have helped and given advice during the preparation of this book. In particular I would mention Dr. R. V. Gotto and Mr. J. B. Taggart, who critically read the entire manuscript and made many useful comments; also Dr. D. R. Bamford, Miss F. C. M. Craig and Dr. M. W. Steer for their valuable assistance with individual chapters. Finally, I would like to acknowledge the help and encouragement given by my wife, Helen, at all stages during its production.

A.F.

Contents

CHAPTER ONE

INTRODUCTION TO SYSTEMATICS

SYSTEMATICS IS REGARDED BY SOME BIOLOGISTS AS SYNONYMOUS with taxonomy. By others, such as Blackwelder (1967), Mayr (1969*b*) and Simpson (1961), it is used in a comprehensive sense to cover the study of the diversity of organisms and their relationships. In this book, *systematics* is used in the widest sense, including aspects of the study of *evolution* (the process by which the diversity of living organisms is produced) and *taxonomy* (the process by which this variation is arranged into a meaningful and useful order). Both evolution and taxonomy involve a study of variation among organisms. Biochemical systematics is the study of biochemical variation, which here will be restricted to a consideration of nucleic acid and protein macromolecules.

Within the field of systematics, several other terms are used with various meanings by different authors. Adequate definition of all terms would require an entire book, so consideration will be given here only to those necessary for an understanding of later chapters. The term *classification* overlaps with taxonomy, but generally it is restricted to that part of taxonomy which involves the arrangement of organisms into groups in an hierarchical system (likewise called a classification) on the basis of their relationships.

Populations and species

If all the organisms living in a particular area are studied, it is found that, although individuals, may be unique if examined closely enough, they occur in groups of similar individuals with a number of features in common. Such groups of recognizably similar individuals constitute *populations*. In outbreeding sexual organisms, the individuals in a population interbreed among themselves, exchanging genes freely, and are said to share a common gene pool. To distinguish it from other uses of the term 'population', this is sometimes referred to as a Mendelian population (or *gamodeme* in botanical useage) and it is with this connotation that the term is used subsequently.

The population, not the individuals comprising it, is the natural unit of evolution and is therefore the basic unit used in systematic studies. A *species* is a group of one or more populations which is given a formal distinguishing name. If the populations, or groups of populations, are sufficiently distinct, they may be given sub-specific names.

In general, the individuals that comprise a species are capable of inter-breeding to produce viable and fertile offspring but, due to the geographical spread of the constituent populations, they may not have the opportunity to do so. Individuals of one species are, as a rule, reproductively isolated from individuals of all other species. However, occasional or localized inter-breeding (or inter-breeding in captivity) does not negate the specific status of the participants. Obviously if two populations are *sympatric* (live in the same area) without inter-breeding, they are separate species. If they are *allopatric* (live in mutually exclusive areas with an unoccupied zone between), evidence of reproductive isolation is not available, and delimitation of species is then generally based on their possession of certain features which are not present in others. Normally these features are aspects of internal or external structure (i.e. morphology), and such species are referred to as *morphospecies*, as distinct from *biospecies*, which are described on the basis of the biological species concept of reproductive isolation. Since the biological species concept does not apply to asexual, parthenogenetic, or self-fertilizing organisms, here again morphospecies are the rule. The term *agamospecies* has been applied to all species that reproduce asexually. Several terms have been proposed for species delimited by biochemical characters, but these terms (such as physiological species) are misleading and superfluous.

Evolution of diversity—speciation

Classifications are necessary because of the immense variability of organisms. This diversity occurs both within populations, and among populations and species. Variation can either be genetically or environmentally induced, or can be the product of a combination of these two forces.

Genetic variation is produced in organisms by mutation and it is this variation which is the basis of all evolution. In most species, more offspring are produced than can survive if the species numbers are to remain stable. For example, individual female fish such as the cod *Gadus morhua* produce upwards of ten million eggs each year, but only two offspring need survive to replace their parents over their reproductive lifespan. Since individual offspring differ in their genetic make-up, they will differ in their ability to cope with the environment in which they find themselves. Those which survive are those which are genetically best adapted, and thus they in turn will produce the next generation. *Natural selection* is the name given to this process, whereby the best-adapted individuals survive to reproduce. *Random processes* may also be important in deciding which individuals survive. The relative importance of natural selection and random events has been the subject of debate among biologists for many decades.

Due to mutation, over-production of offspring, natural selection and stochastic events, the genetic composition of a population will change gradually, i.e. it will evolve. Over the generations there is a constant refinement of the genetic composition of the population, as the genes selected for are those best adapted to the conditions in which the population lives. However, the environment is constantly changing, and a stable situation is not usually attained for a long time, since new adaptations are frequently necessary. Over a period of many thousands of years, a species will change in its genetic make-up and consequently in its structural and other features. After a time it may have changed sufficiently for the taxonomist to regard it as a different species. This process, which is a change within a single lineage, is called *anagenic* or *phyletic speciation* (figure 1.1).

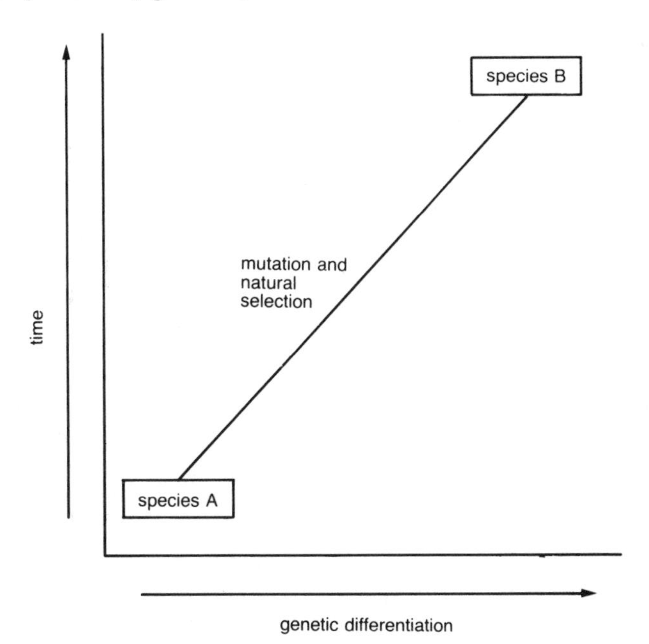

Figure 1.1. Diagram illustrating the process of anagenic speciation.

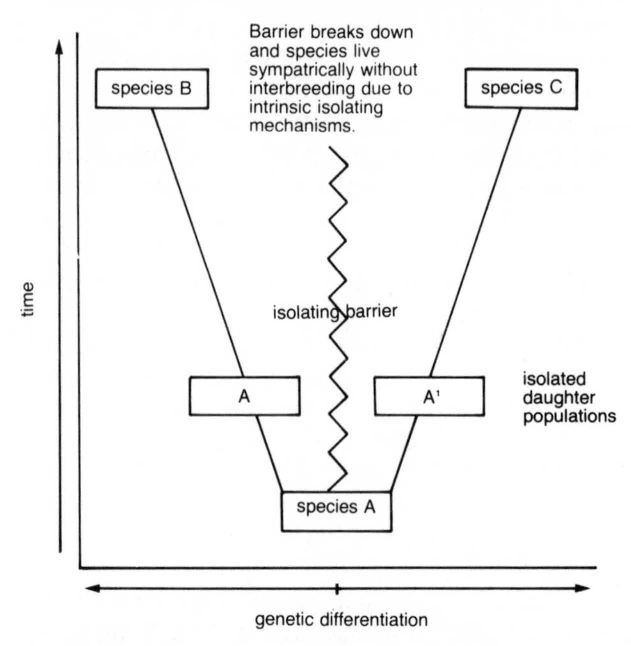

Figure 1.2. The process of allopatric cladogenic speciation. When two populations of a species become isolated, they diverge genetically due to independent mutation, natural selection and genetic drift. After a time, even if the barrier breaks down, they will not interbreed. In time, these species may be split into isolates which will likewise diverge.

More important, however, is the process of *cladogenic or dichotomous speciation* (figure 1.2) since this produces an increase in the number of species. Populations of a species which are isolated from each other will acquire different random mutations and will diverge genetically. If they are isolated for a sufficiently long period, they may acquire sufficient differences to be regarded as separate species. It is this continuous nature of allopatric speciation that makes delimitation of some species so difficult. An appreciation of the uninterrupted sequence of events involved in most speciation will also show the futility of many of the arguments pertaining to the 'species problem'. At one end of the process there are populations which have acquired few genetic differences and are obviously of the same species. At the other end, the populations have diverged greatly and are no longer capable of inter-breeding—clearly these are now separate species. However, between these two extremes there are many cases where it is difficult to decide if two populations belong to the same species or not. Some populations may be on the way to becoming separate species and have acquired some, but not yet all, of the features of distinct species. If the barrier separating them breaks down, they may inter-breed as a single population. It is not possible to establish a firm boundary below which the populations are conspecific (= same species) and above which they are different species.

While geographical isolation of populations, coupled with the gradual accumulation of mutations, is an important mechanism of cladogenic speciation, it is not the only one. Chromosomal and other changes resulting in speciation are discussed in chapters seven and nine.

Isolating mechanisms

In the course of evolution, each species acquires a unique set of genes which adapt it to its particular mode of life. If species were to inter-breed, these unique adaptations would become diluted or lost. As might be expected, therefore, most species possess one or more mechanisms which ensure reproductive isolation. These mechanisms may be pre-mating features which prevent inter-specific mating (e.g. mechanical, ethological, seasonal and habitat isolation) or post-mating, which reduce the success of inter-specific crosses (e.g. gametic or zygotic mortality, hybrid inviability or sterility).

There is little exact information on how isolating mechanisms arise, but there may be two main ways (Ayala, 1975). Isolation may be produced as a direct result of genetic differentiation of geographically isolated, diverging populations. This is probably the origin of post-mating mechanisms, especially where the zygote fails to develop normally due to the incompatibility of the two different gametic genomes. If the barrier isolating two populations breaks down, permitting inter-breeding, then either the two populations will fuse into one or they will remain distinct. In the latter case, if matings between individuals of the different populations produce offspring with reduced fertility or viability, natural selection will favour any mutations which promote matings between individuals of the same population. Such pre-mating isolating mechanisms are reinforced when the two populations are sympatric.

Hybridization

Reproductive isolation may break down even between good species. However, the viability and fertility of the hybrids produced may be reduced or absent. Rarely, there is a complete breakdown of isolation resulting in extensive hybridization, introgression, and eventual fusion of the gene of the two populations. Hybrids may be produced in a localized zone between species with a *parapatric* distribution (occupying contiguous geographical areas with a very narrow zone of overlap). For example, hybrids between the hooded crow *Corvus cornix* and the carrion crow *C.*

corone are restricted to a zone about 80 km wide throughout Europe between the ranges of the two species. Also of considerable interest are the two species of towhees studied in Mexico by Sibley (1954). The red-eyed towhee *Pipilo erythrophthalmus* and the collared towhee *P. ocai* live sympatrically in some parts of their range without inter-breeding, but in other parts, where they are forced to share the same habitat, a large proportion of hybrids is found. Similarly, in certain parts of California only 'pure' oaks of the species *Quercus douglasii* and *Q. turbinella* are found, while in others inter-breeding takes place (Tucker, 1952).

Classification

By cladogenic speciation there has evolved an immense diversity of organisms which the taxonomist must fit into a scheme of classification. The closer the relationship of organisms, the nearer they are placed in a classification. Relationship can imply derivation from a common ancestor, or may relate to overall similarity in various features.

The features or attributes of organisms that are used to establish relationships are called *characters*. All types of information are potentially of value as systematic characters. Each of these exists in one or more character states. These states may be simply presence or absence, or may be much more complex, as is the case with many meristic characters. The characters used in systematic studies are derived from observations of morphology, physiology, biochemistry, ecology, genetics, behaviour, parasitology and geographical distribution of organisms. Until recently, the chief source of information derived from morphological investigation, because structural features can be examined in preserved museum specimens—the main material with which the systematist had to work in the past.

Considerable debate has taken place among systematists as to whether or not classifications should reflect the evolutionary history (phylogeny) of organisms. Thus should taxa be placed in the same group if they have evolved from a common ancestor, irrespective of their present-day similarities? (A *taxon* is a group of actual organisms assigned to a definite taxonomic category, e.g. *Homo sapiens* is a taxon of species rank which is part of a larger ordinal-level taxon, the Primates.)

Three main schools of systematics have developed, each one claiming to be more objective and of more practical value than the other two.

(1) *Phenetic systematics.* Pheneticists contend that classifications should be based on overall similarities among living organisms. In general this approach is equated with numerical taxonomy (e.g. Sneath and Sokal, 1973) and involves examination of all possible characters and

calculation of average similarities. In this approach all characters are assumed to be of equal importance, and no account is taken of genealogical relationships.

(2) *Cladistic (phylogenetic) systematics.* Exponents of this approach (e.g. Hennig, 1966) use only cladistic relationships as a basis for constructing classifications, i.e. their emphasis is on the sequence of cladistic splitting of lineages in the origin of taxa. The rate or amount of adaptive change subsequent to the splitting of phyletic lines is not considered. Taxa in cladistic classifications are required to be monophyletic (*sensu* Hennig), i.e. all must have arisen by cladogenic speciation from a common ancestral species. (Other systematists use the term *monophyletic* in a different way, meaning origin from a common ancestral taxon, even if by more than one lineage, and suggest the term *holophyletic* for Hennig's concept). Since birds and crocodiles (excluding all other living reptiles) are believed to be derived from a common ancestral archosaur, the cladists would put birds and crocodiles together in a taxon. In a truly cladistic classification, recency of common ancestry is used as a criterion for ranking of taxa. Thus taxa that evolved earlier are given a higher rank than those of later origin, and the same rank is given to sister groups. Taxa that originated from a cladogenic split in the Pre-Cambrian are ranked as phyla; those that originated in the Miocene as genera. Thus the coelacanth originated before either birds or mammals, and should be given a higher rank than either of these latter groups. Similarly crocodiles and birds should be of equal rank.

(3) *Evolutionary systematics.* Evolutionary systematists (e.g. Simpson, 1961; Mayr, 1969b) use the order of origin of lineages, and also take into account the amount and nature of evolutionary change which occurs after cladogenesis. This school thus combines elements of both phenetic and cladistic classifications. In determining relationships, all characters are not regarded as of equal value. Major adaptive shifts are regarded as more informative, and are weighted accordingly.

The evolutionary taxonomist believes that an approach which superimposes a carefully weighted phenetic analysis on a preceding cladistic analysis is better able to establish degree of relationship than either a pure cladistic or an unweighted phenetic approach (Mayr, 1974).

The major adaptive changes which have taken place in the bird lineage serve to separate birds from crocodiles in spite of common ancestry. The combination of phenetic and cladistic information is to some degree subjective, in contrast to the more formalistic methods of the other two schools. Problems also arise in trying to express both types of information adequately in a single classification.

In some cases, classifications based on phenetic similarities may also reflect the phylogeny of the taxa, those which are most similar probably

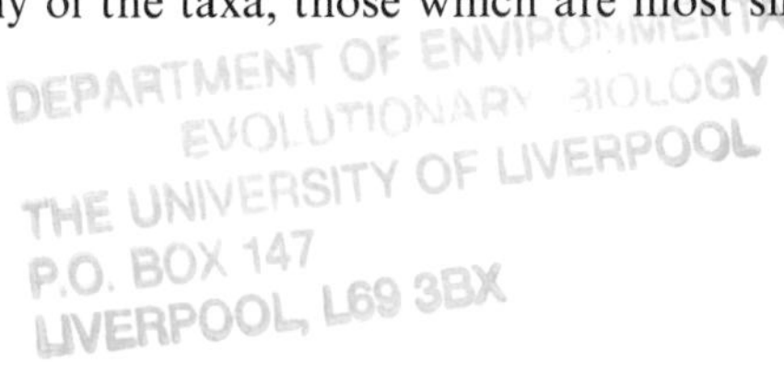

having evolved from a common ancestor. In other cases this is not so, due to convergent evolution, i.e. the acquisition of similar features by two unrelated species due to adaptation for life in the same environment. For example, fish and whales have a similar body shape as this is the most efficient one for life in water. An examination of other features show, however, that they are very different. Pheneticists argue that if sufficient characters are used (60+) then the contribution of convergent ones will be minimal and unimportant. Even so, the possibility exists that morphologically similar but unrelated organisms will be placed in the same group, and related but morphologically divergent ones in different groups.

As will be seen in later chapters, biochemical studies of macromolecules can disclose both phenetic and phylogenetic information, depending on the type of study employed. One of the major difficulties in any reconstruction of phylogeny is to discriminate between primitive or ancestral (plesiomorphic) and derivative (apomorphic) character states. Since molecular evolution is a continual process and produces, in the main, differentiation, the differences among extant species are primarily derived character states.

Graphical representation of classifications

A classification can be represented in graphical form as a tree-like dichotomous branching graph or *dendrogram*. A dendrogram produced from phenetic information is called a *phenogram*, that from cladistic information a *cladogram*, and that embodying both phenetic and phylogenetic data is a *phylogram* or *phylogenetic tree*. A phenogram (figure 1.3) shows the present-day similarities of the group but does not indicate the probable lines of descent. A cladogram (figure 1.4) shows the sequence of origin of the various lineages and indicates the times at which the various cladogenic events have taken place. A phylogram indicates the cladistic branching and, by the angles and lengths of the branches, the relative amount of change that has taken place in each branch (figure 1.5).

Hierarchical system of classification

Those species which show the closest phenetic or phylogenetic relationship are grouped together into larger, more inclusive groups or genera. A number of genera are in turn grouped into families and so on, in increasingly inclusive categories: orders, classes, phyla. Subdivision of

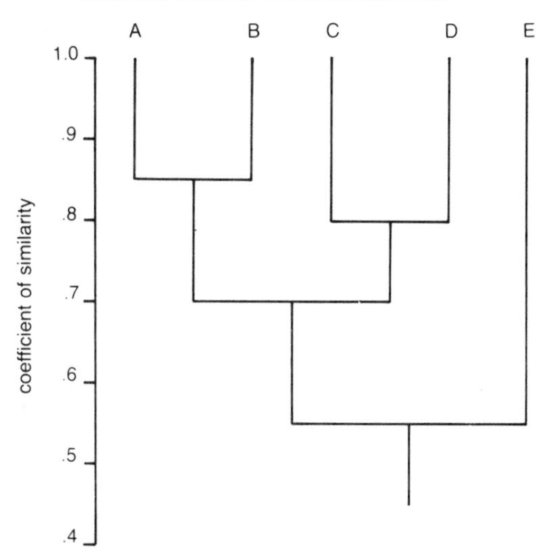

Figure 1.3. A phenogram showing the phenetic relationships of five taxa. The ordinate scale indicates the degree of similarity. The scale of the abscissa is arbitary. The horizontal lines do not indicate cladistic branching, but only the similarity between the two taxa which they join. Taxa A and B are phenetically the closest.

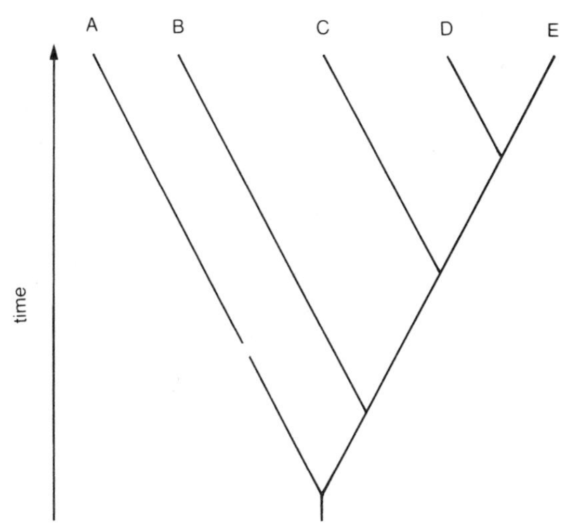

Figure 1.4. A cladogram showing the sequence of origin of five taxa. In a cladistic classification, D and E would form a low-rank taxon (e.g. genus); C, D and E, one of higher rank (e.g. family), and so on.

the main groupings results in some twenty different categories being used in the classification of organisms, but these fit into three natural groups:

(1) Categories for distinguishing populations within a species (intraspecific taxonomy).
(2) The species category.
(3) Categories for groups of species (higher-category taxonomy).

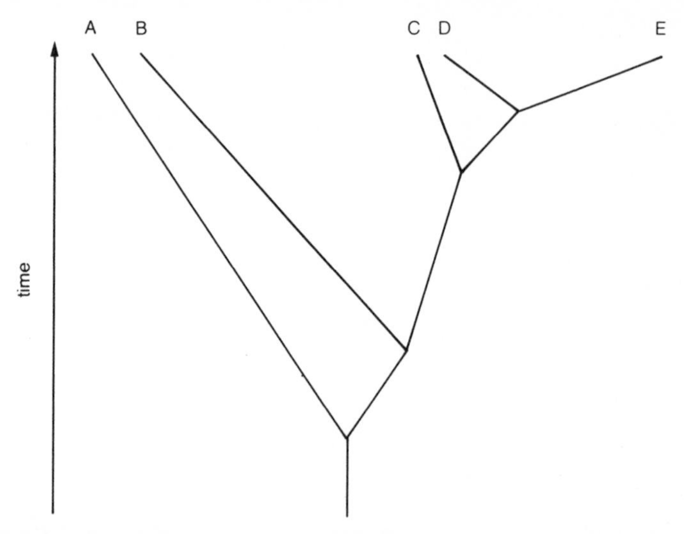

Figure 1.5. A phylogram showing the phylogenetic relationships of five taxa. Three types of information are displayed: the sequence and time of branching (ordinate); present-day similarity (abscissa); the rate of evolution (angle of divergence—the slower a lineage evolves the more vertical the line). Thus taxa A and B, although the product of an ancient cladogenic event, have evolved in parallel and with a slight degree of convergence. Taxon E, while sharing a recent common ancestry with D, has evolved at a rapid rate and is phenetically distinct.

Variation within the species

For the taxonomic delimitation of any category, character states are needed which are unique to each of the constituent groups. Thus, for the separation of species, features must be possessed by all members of the species, and not by any other species. For most features there is variation among the individuals which make up a single population among the populations which make up the species. How to distinguish the variation within the species from the differences between species has always been one of the chief difficulties in systematics. Two individuals, although belonging to the same species, may differ because of individual genetic variation, differences in age, sex, food, stage of life cycle, body form, habitat, disease, accidental damage and parasitism. Many errors in classifications have been made because of failure to recognize this intra-specific variation. Thus stages in the life cycle, different sexes, and individuals feeding on different foods have in the past been described as separate species. The systematist tries to obviate this problem by examining a series of specimens rather than a single representative, in the hope of determining the range of variation in the chosen features.

By far the largest source of intra-specific variation is the effect of environmental factors. These can cause temporary, cyclic, or permanent changes in the individual. Many of the morphological features of organisms can change during the lifetime of the individual or be determined during development in response to environmental conditions. For example, in the brown trout (*Salmo trutta*) there is considerable variation in colour, and in many meristic features depending on the water body in which the individual fish is living. When in the sea, the body colouration is silvery white but, when the trout enters fresh water, it takes on the colouration typical of a particular area, which may range from a dorsal colour of bluish-grey or olive, through yellows and browns, to nearly black. The spots on the sides may be red or brown, and they may vary greatly in size, shape, number and distribution. Trout of common parentage, reared in a hatchery and stocked in different waters, soon assume the colours typical of these waters. In addition, the numbers of vertebrae and fin rays in an individual depend on the temperature of development, being maximal at intermediate temperatures (figure 5.3). Numerous other examples of environmental modulation of morphological characters in both plants and animals have been described.

Environmental and other non-genetic influences bedevil the use of morphological characters in systematics and lead to diverse interpretations of relationships. There is, of course, genetic variation within species, each individual (with the exception of monozygotic twins) being genetically unique. In some cases, such genetic variation is discontinuous and, for a particular character, can be fitted into a small number of definite types, whereas environmentally induced variation is generally continuous and can only be assessed subjectively. The genotype does not change during the lifetime of the individual, although there may be differential expression of some genes.

It was noted earlier that both evolution and taxonomy are studies of variation. This might now be qualified to genetic variation, evolution being the study of genetic changes, and taxonomy the study of genetic similarities and differences. Ideally, in order to determine variation among organisms, we should obtain the actual arrangement of the genetic material. Such information would be free from environmental influences but would still result largely in a phenetic classification. This is due to the action of selection on the genomes, resulting in convergent similarity of unrelated forms and divergence of related ones. For many purposes, a complete genetic catalogue of each population would be the ideal. In practice, the names given to different populations, species, etc., or the ways in which they are classified do not matter, as long as the extents of genetic similarities and differences are known. Biochemical techniques make a complete genetic characterization more of a practical proposition than hitherto, and the use of these techniques in examining genetic variation will be examined in subsequent chapters.

The role of systematics in modern biology

Systematics is the most inclusive area of biology, since it takes into account all that is known about an organism. Conversely, the knowledge of an organism's classification immediately gives a tremendous amount of information about it. A classification provides the framework for the organization of information on the enormous diversity of organisms. Without this framework no other biological studies can be carried out in a rational manner. In certain groups such as birds, most of the living species in the world have been described, but even in this class considerable improvements are possible in the higher categories to better reflect phylogeny. In many invertebrate groups, numerous species still remain to be described and classified accurately.

Proper identification of all the species involved is a prerequisite for most ecological studies, and this identification is dependent upon an adequate classification being available. In groups such as the chironomid midges, some ecological work is reduced in value because of mis- or non-identification. This applies especially to the use of chironomids as biological or environmental indicators (Bryce and Hobart, 1972). For example, mesotrophic lakes have *Chironomus cingulatus* as the predominant chironomid, while eutrophic lakes are characterized by *C. plumosus*. As larvae these two species are morphologically inseparable, which means that they are valueless as indicator species. However, biochemical techniques can readily separate them (p. 105).

Soundly based taxonomy is necessary also for the experimental biologist. There are many genera with two, three or more very similar species. Such species often differ more markedly in their physiology, cytology, behaviour, etc., than they do in their morphological characters. Reliable classification at the species level is thus necessary, so that biologists can avoid reaching different conclusions about the properties of a particular 'species', because unintentionally they are working with different species.

Of particular importance is intra-specific systematics, and it is in this area that most work is required for many organisms. A knowledge of population taxonomy is necessary for the rational exploitation and management of natural populations of food-resource animals and plants. A documentation of the genetic diversity within a species is vital for the conservation of unique genotypes for potential future use in breeding programmes. The characterization of disease-carrying and resistant strains, as well as of genotypes correlated with economically desirable traits, is of value in medical and agricultural work.

With the realization that a large amount of genetic variation exists within populations and species, there is an increasing need for methods to identify specific genotypes. Since the genotype of an individual determines its structure, physiology, behaviour, etc., workers in these disciplines need to be certain of the genetic composition of their experimental organisms. Two genotypes within a population of the same species may differ greatly in some aspect of their biology, while two individuals of different species, if they have the same ancestral genotype for a particular gene, may be identical. Figure 4.8 shows the enzymatic activities of three esterase genotypes of the fish *Catostomus clarki*, as determined by Koehn (1970) (p. 65). A physiologist examining the relationship between temperature and activity for this enzyme would come to very different conclusions from different individuals. Traditionally, biologists have attempted to allow for genetic variation within a species by statistical treatment of results. In this case such a treatment would obscure the true nature of the enzyme's activity relative to temperature. Thus each genotype category would need to be analysed separately. Even inbred 'pure' lines of laboratory rats have been shown to contain considerable genetic variation (p. 59).

History of the biochemical approach

The advantages of using biochemical characters to refine aspects of systematics were recognized at the beginning of the century by Nuttall (1901) and by Bateson (1913). Much of the work in the first 40 years of this century was concerned with immunologically determined similarities and differences. In the 'forties and early 'fifties, other techniques—such as moving boundary and paper electrophoresis—were used on a small scale, but the techniques were too slow and yielded too poor resolution for detailed comparative work.

The significant application of biochemical characters had to await the development of more powerful techniques. In the 'fifties, with the development of gel electrophoresis and the discovery that genetic information is carried from generation to generation encoded in the nucleotides of the DNA, biochemical systematics was given a renewed stimulus. The late 'sixties and early 'seventies saw a large upsurge in comparative biochemical studies, many of which were directly or indirectly of systematic interest. Much of the early work suffered from a lack of recognition and appreciation of intra-population and intra-specific variation, with systematic conclusions being based on one or a few individuals. Biochemical techniques are now being applied to systematic problems of all types and in many different organisms.

The term *chemosystematics* is sometimes used synonymously with *biochemical systematics*, although the former tends to be favoured by botanists and the latter by zoologists. In part this stems from the use by botanists, in the past, of low-molecular-weight chemicals—the study of which is the realm of the chemist—while zoologists have concentrated on proteins and nucleic acids—the concern of the biochemist.

CHAPTER TWO

RATIONALE FOR THE BIOCHEMICAL APPROACH

AS DISCUSSED IN CHAPTER ONE, THE STUDY OF EVOLUTION AND TAXONOMY involves an investigation of the changes and variation in the genetic constitution of organisms. How then are genetic similarities and differences examined? There are two principal approaches to this problem: the structures of genes can be examined either directly, or indirectly through their products.

The genetic material

In order to appreciate the types of approach that can be used, a brief outline of the structure, functioning and modification of the genetic material is appropriate. In most organisms the genetic message is carried from generation to generation encoded in molecules of deoxyribonucleic acid (DNA) which is a large polymer of individual units called *nucleotides*. In some primitive prokaryotic (without nuclei) organisms, e.g. bacteria, DNA is present as a single chain molecule. In eukaryotic (with nuclei) organisms the DNA molecule is in the form of a double helix which forms the axis of large nuclear structures, the chromosomes of the cell. In some viruses ribonucleic acid (RNA) rather than DNA is the hereditary material. In plants (and some animals) extra-chromosomal DNA is found in the mitochondria and chloroplasts, and these genes are maternally inherited through the cytoplasm of the ovum.

Each DNA nucleotide consists of a nitrogen-containing base, a deoxyribose sugar and a phosphate group. The double-helix molecule is rather like a twisted rope ladder in which the phosphate and sugar components alternate to form the sides, while the bases join in pairs to make the cross-bars. Whereas the sides are of uniform structure throughout the length of the molecule, the cross-bars vary, since the bases that form them are of four different types: two purines (adenine and guanine) and two pyrimidines (cytosine and thymine). Their molecular structure is such that adenine (A) always pairs with thymine (T) and cytosine (C) with guanine (G). This pairing constraint ensures the precise replication of the DNA molecule, a pre-requisite for genetic material. The *sequence* of bases along the length of the molecule forms the genetic code.

Structure of proteins

Proteins are composed of one or more polypeptides. Polypeptides are chains of some 20 different α amino acids (table 2.1). However, even for a relatively small molecule of 100 amino acids, a vast number ($20^{100} = 10^{130}$) of different polypeptides is possible. Amino acids are nitrogen-containing molecules with an amino group (NH_2) and a carboxyl group (COOH), and a general formula of:

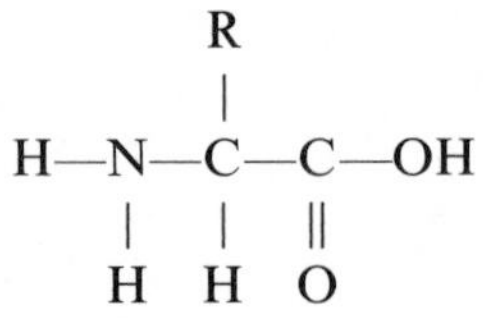

```
        R
        |
  H—N—C—C—OH
    |  |  ||
    H  H  O
```

It is in the composition of the R side-chain group that one amino acid differs from another. The simplest amino acid is glycine, in which R is a hydrogen atom; in alanine R is CH_3, and so on. The specificity of the polypeptides is determined by the order or *sequence* in which the amino acids are present, and not by the absolute amounts. This sequence is known as the *primary structure* of the polypeptide. The individual amino acids are linked through amide or peptide bonds (—CO . NH—) which are formed by the elimination of water between the carboxyl and amino groups of adjacent amino acid residues (figure 2.1). Each polypeptide has a free amino group at one end (N-terminal amino acid) and a free carboxyl group (C-terminal amino acid) at the other end.

Most polypeptides are coiled in a variety of ways by the formation of

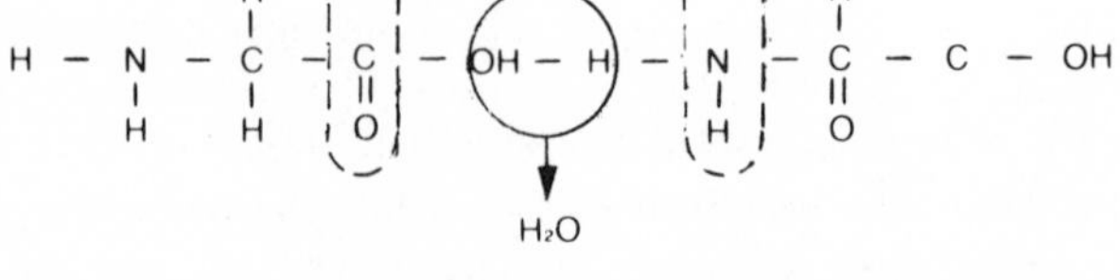

```
                       Peptide bond
          R                                   R'
          |                                   |
  H — N — C — C — OH — H — N — C — C — OH
      |   |   ||             |   ||
      H   H   O              H   O
                   ↓
                  H₂O
```

Figure 2.1. Formation of a peptide bond.

hydrogen bonds between adjacent amino acids. The most common form of coiling is the α-helix which is somewhat like a slightly extended coiled spring. The coiling of a polypeptide is referred to as the *secondary structure*. Disulphide bonds can bridge two cysteine residues in different parts of the coiled chain and this, and other bonding, results in folding of

Table 2.1 Amino acids found in proteins and their genetic codes.

Amino acid	*Abbreviation*	*Single letter abbreviation*	*DNA codon(s)*
Alanine	Ala	A	CGA, CGG, CGT, CGC
Cysteine	Cys	C	ACA, ACG
Aspartic acid	Asp	D	CTA, CTG
Glutamic acid	Glu	E	CTT, CTC
Phenylalanine	Phe	F	AAA, AAG
Glycine	Gly	G	CCA, CCG, CCT, CCC
Histidine	His	H	GTA, GTG
Isoleucine	Ile	I	TAA, TAG, TAT
Lysine	Lys	K	TTT, TTC
Leucine	Leu	L	GAA, GAG, GAT, GAC, AAT, AAC
Methionine	Met	M	TAC
Asparagine	Asn	N	TTA, TTG
Proline	Pro	P	GGA, GGG, GGT, GGC
Glutamine	Gln	Q	GTT, GTC
Arginine	Arg	R	TCA, TCC, GCA, GCG, GCT, GCC
Serine	Ser	S	AGA, AGG, AGT, AGC, TCA, TCG
Threonine	Thr	T	TGA, TGG, TGT, TGC
Valine	Val	V	CAA, CAG, CAT, CAC
Tryptophan	Trp	W	ACC
Tyrosine	Tyr	Y	ATA, ATG
Aspartic acid or asparagine, not distinguished		B[a]	
Glutamic acid or glutamine		Z[a]	
Undetermined		X[a]	

[a] Additional abbreviations used in listing amino acid sequences.

the molecule. The overall folding of the polypeptide chain brings into proximity amino acids far apart in the primary structure, and is known as the *tertiary structure*. Many proteins are made up of more than one polypeptide chain, in most cases two or four (table 4.1). The combination of subunits into a specific protein is the *quaternary structure*.

Changes in temperature, pH, ionic strength, etc., may break down the quaternary or tertiary structure of a protein. Such denaturation generally results in the loss of enzymatic or other biological activity of the protein. There is considerable variation among proteins in the ease with which denaturation occurs.

The R groups of some amino acids are acidic (aspartic acid, glutamic acid) or basic (histidine, lysine, arginine). These acidic or basic groups can exist in uncharged or charged form, depending on the pH of the environment. Under acid conditions, aspartic acid and glutamic acid have an uncharged carboxyl group, whereas histidine, lysine and arginine are protonated and carry a positive charge. Under basic conditions, the carboxyl group of aspartic acid and glutamic acid will be ionized and impart a negative charge, while histidine, lysine and arginine will be uncharged.

Since some of the constituent amino acids are charged, this means that the overall protein will be charged, and this charge will be pH-dependent. At a certain pH or isoelectric point (pI) the protein will have an equal number of positively and negatively charged groups and be electrically neutral. Since proteins differ in their amino acid composition, they will vary in the magnitude and polarity of their electric charge at a given pH.

Plasticity of protein structure

An important feature of proteins is that different primary structures can still give proteins with the same structural or enzymatic function. The enzyme cytochrome c is found in all organisms with aerobic metabolism, and its amino acid sequence has been determined in some 50 organisms ranging from bacteria and yeast to man. It consists of a single polypeptide chain of from 103 to 112 amino acids. Only 20 of the amino acid positions are invariant in all organisms but, in spite of this, the enzyme has essentially the same function throughout. Similar, though less extensive, differences have been found in other proteins when examined in a range of species.

Each protein seems to have a certain number of amino acid positions which must remain invariant if the function is to be retained. For enzymes these are presumably the amino acids involved in forming the active sites. In other parts of the molecule a number of different amino acids may be equally suitable. However, even though two proteins with different primary structures appear to carry out the same function, they may differ in their kinetic or other properties. This may alter the efficiency with which they carry out their function, and may be important in adaptation to diverse environmental conditions. This will be discussed further in chapter 4.

The genetic code

One of the functions of genes is to assign amino acids to their correct positions in the sequences of polypeptides. Since there are four bases but

20 amino acids, single bases cannot code for amino acids. A combination of 2 is not sufficient ($4^2 = 16$), but triplet combinations give more than enough ($4^3 = 64$). Each triplet set of bases which codes for a single amino acid is called a *codon*. Since there are more possible codes than there are amino acids, there is more than one codon for most (table 2.1). In this sense, the genetic code is said to be *redundant*. Three codes also denote the termination of the polypeptide chain.

The sequence of codons on the DNA strand specifies the sequence of amino acids in the protein coded for by that section of DNA. Each group of codons or genes codes for a single polypeptide chain, which in some cases equates with a single protein.

Genotype and phenotype

The genetic information contained in the DNA controls the structure and functioning of the organism through the processes of transcription and translation. Broadly speaking, genes may be classified into two types: *structural* genes are those that code for proteins; all other genes are *regulatory*, i.e. they govern the production of proteins by structural genes (rate, quantity, timing, etc.) and the aggregation of these proteins to form the entire organism. It is at this later stage that environmental effects come into operation and mould the genetic information into the final phenotype, i.e. the morphology, physiology and behaviour of the organism. Because of this interaction between genes and environment, the genotype does not unambiguously specify the phenotype, but only the range of phenotypes which may be produced. The phenotype of the individual in many cases changes during its life, while the genotype remains constant except for mutation. However, the structure of specific proteins is a primary phenotype which is free of environmental influence and remains constant throughout the individual's life.

Arrangement of DNA

The amount of DNA which is present in organisms varies greatly, even among those which are closely related. Vertebrates have some 300–10 000 times more DNA per gamete than does a bacterial cell, even though 93% of known enzyme activities are present in both prokaryotes and mammals. The genome size ranges from about 1×10^8 nucleotide pairs per haploid set for *Drosophila melanogaster* to 8×10^{10} for the urodele amphibian *Amphiuma*. Nucleotide sequences in DNA are of three principal types:

1. Unique sequences which are present as a single copy.

2. Sequences repeated a few times, possibly to code for proteins required in large amounts.
3. Sequences repeated hundreds or thousands of times. This highly repetitive DNA may play a role in the regulation of structural genes.

Most structural genes are present as single copy DNA, but the majority of single copy DNA is not made up of structural genes (Britten and Davidson, 1976). The exact number of structural genes which are present is unknown but estimates range from 5000 for bacteria up to 100 000 for mammals, and this may involve less than 1% of the total DNA.

Transcription and translation of the genetic message

The process of transcription takes place in the nucleus except for mitochondrial and chloroplast genes. The sequences of bases on one of the DNA strands which make up a gene is transcribed with a high degree of accuracy into a complementary messenger RNA sequence. In this transcription the same base pairing takes place as in the replication of DNA, except that thymine is replaced by uracil in RNA. This messenger RNA then likewise carries the information for the amino-acid sequence in polypeptides. The process by which the information encoded in the messenger RNA is used to direct the production of a polypeptide is called *translation.* Translation takes place in the cytoplasm and is brought about by ribosomes, transfer RNA molecules and several enzymes.

Genetic variation

The process of evolution depends on the occurrence of genetic variation. If DNA replication was always perfect, life could not have evolved and diversified. In general, genetic changes fall into one of three categories:

point mutations involving changes in a single codon;
chromosomal aberrations in which large sections of the chromosome undergo change;
changes in chromosome number.

Frequently the change in a codon entails only the substitution of one base for another in a single base pair within the DNA molecule. Since there is more than one code for most amino acids, the new codon may specify the same amino acid as before. If, for example, the DNA codon ATA is changed by a point mutation to ATG, no change would occur in the protein resulting from this gene, as both sequences specify the amino acid tyrosine. On the other hand, the new codon may determine a different amino acid, e.g. ATA coding tyrosine may be altered by a single

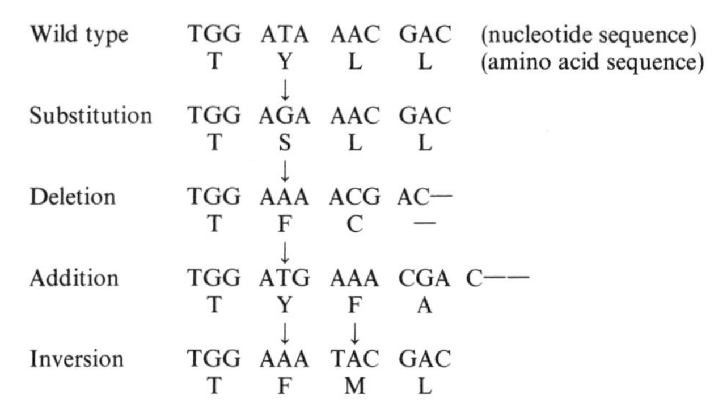

Figure 2.2. An illustration of the four basic types of single nucleotide mutation. The base sequence is given in units of codons or nucleotide triplets in order to illustrate how the amino acids coded for are changed by the nucleotide changes. The single-letter amino acid codes are given in table 2.1 (after Nei, 1975).

base change to AGA, which codes serine (figure 2.2). Finally, the new codon may signify punctuation instead of an amino acid. Change in the sequence CTT (glutamic acid) could result in the new codon ATT, which indicates the termination of the polypeptide chain. Consequently, the protein molecule would be shortened, possibly to such an extent as to be non-functional, or else it might conceivably function in a completely different manner.

Instead of base substitutions, point mutations can result in the addition or loss of a base. Because the transcription proceeds from a fixed starting point and is 'read' in groups of three bases, the gain or removal of a base causes a shift in the 'reading frame' of the codons, beginning at the site of the mutation and continuing to the extreme end of the gene. This type of frame-shift mutation obviously produces a very drastic change in the amino-acid sequence of the protein produced and almost always leads to the production of completely non-functional proteins.

Point mutations occur spontaneously due to naturally occurring replication errors. The occurrence of point mutations may be increased by exposure to high-frequency radiations and certain chemicals. The rates of natural mutation vary from organism to organism, and from gene to gene within the same organism. In viruses, bacteria and unicellular organisms, the range per gene per cell division is from 10^{-6} to 10^{-9}. In higher organisms, observed mutation rates are of the order of 10^{-4} to 10^{-6} per gene per gamete. This means that if a higher organism has some 50 000 genes, at an average mutation rate of 10^{-5} each individual would have $2 \times 50\,000 \times 10^{-5} = 1$ new mutation. Even for a given gene, the incidence of new mutations is high if the entire species is considered. For each million individuals in the species,

$2 \times 10^6 \times 10^{-5} = 20$ new mutations would arise for each gene at every generation. The potential of point mutation to generate new genetic variation is substantial, considering the number of genes and the number of individuals which make up populations.

While mutation is the ultimate source of all variation upon which evolution is based, recombination is responsible for the shuffling of genes and producing new combinations in sexually reproducing organisms. Since it is the overall genotype of the individual which is subject to selection, a few mutations can produce a vast number of different combinations. Recombination is of two principal types:

1. *Independent segregation* of chromosomes at meiosis, which results in the random mixing of chromosomes from the two parents, producing all possible combinations in the zygote. For example, if there are three pairs of homologous chromosomes (AA',BB',CC'), gametes with eight different combinations of these chromosomes can be produced (ABC, ABC', AB'C, AB'C', A'BC, A'BC', A'B'C, A'B'C'). In such zygotes the number of different chromosome combinations produced from the union of gametes from two parents is $8^2 = 64$. In man, with 23 pairs of chromosomes, there are $2^{23} = 8\,388\,608$ different gametic chromosomal combinations.

2. *Crossing over*, whereby exchange of genes takes place between homologous chromosomes, and *translocation*, where a section of one chromosome is transferred to a non-homologous one.

Chromosome mutation may involve duplication where genes are repeated during replication. Other changes are inversion, where a segment of the chromosome becomes broken off, but is rejoined in an inverted position, thus reversing the gene sequence in that part and deletion where part of the chromosome is lost. Changes in chromosome number may occur as a result of fusion or fission of particular chromosomes, or as a result of loss or gain of one or more chromosomes (aneuploidy). Complete new sets of chromosomes may be added. Most higher organisms have chromosomes occurring in homologous pairs, i.e. they have a double or diploid set. Some species are polyploid and have acquired triploid, tetraploid or more sets, while others have only a single haploid complement.

Gene duplication

Gene duplication, either as individual genes or as sets through polyploidy, brings about an increase in the total amount of DNA and is thought to have been very important in evolution. Duplicated genes can

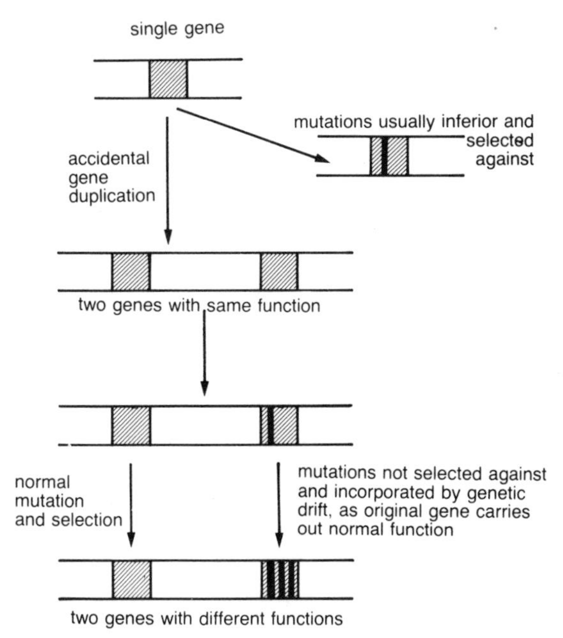

Figure 2.3. An illustration of the importance of gene duplication in evolution (after Dickerson and Geis, 1969).

diverge and acquire new functions (figure 2.3). In the case of vertebrate globins it seems likely that the five genes which code for the α, β, γ and δ haemoglobin polypeptides and myoglobin found in man have arisen by duplication from a common ancestral gene. The first duplication occurred some 500 million years ago, and led to myoglobin and the ancestral vertebrate haemoglobin gene. Myoglobin consists of a single polypeptide chain (monomer), whereas vertebrate haemoglobin (with the exception of that found in the Agnatha) is a tetramer of two polypeptide chains of one type and two of another. In the Agnatha a single gene only is present, whereas in the Teleostei two genes (α and β) are found, so duplication must have occurred somewhere in the early history of the bony fishes. The original haemoglobin gene is thought to have been β-like. The γ gene is found only in therian mammals, and duplication leading to its divergence from the β gene must have occurred after the divergence of the prototherians from the line leading to the marsupials and higher mammals. The δ gene is found only in the hominoids and not in the lower primates and is thus of more recent divergence (figure 2.4).

The duplications of the α-β and β-γ genes were accompanied or followed by chromosomal translocations since, in man at any rate, they are found on different chromosomes. During the evolution of the different globin genes, additions and deletions, as well as substitutions, have taken place. Thus human myoglobin consists of 153 amino acids,

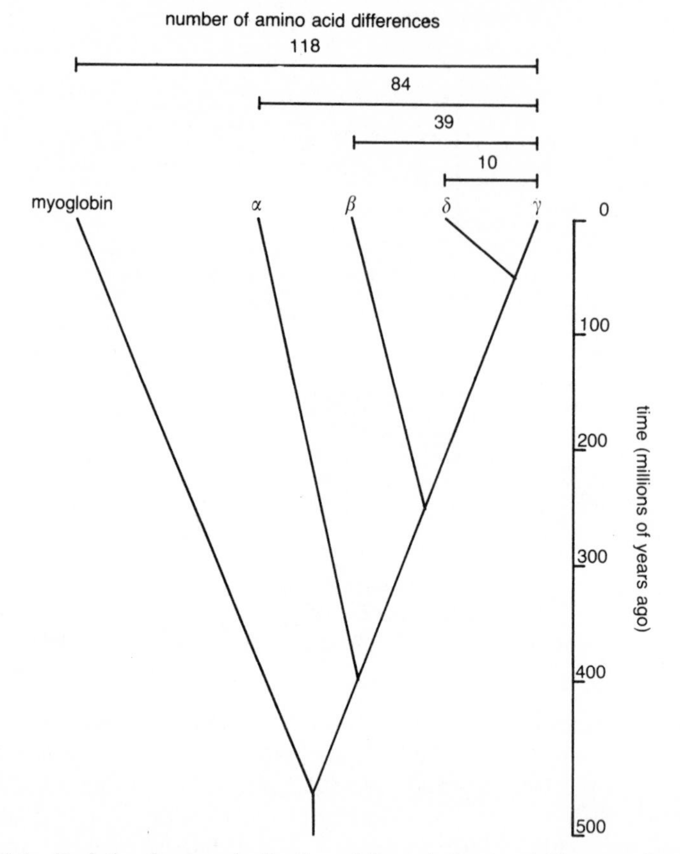

Figure 2.4. Evolution, by gene duplication, of the main human globin genes. See text for further details. The divergence times are from various sources and the amino acid differences are from Dayhoff (1976).

the α chain of 141 amino acids and the β, γ and δ polypeptides are each 146 amino acids in length.

A similar process of gene duplication has probably given rise to the multiple lactate dehydrogenase genes found in higher vertebrates. The Agnatha have a single lactate dehydrogenase gene, but bony fishes and higher vertebrates have at least three. MacIntyre (1976) has reviewed the occurrence and evolutionary importance of duplicate genes. Further aspects will be discussed in chapters 7 and 8.

Homology

Proteins coded for by structural genes which have evolved from a common ancestral gene are said to be *homologous*. In general, proteins which serve the same function in different taxa are homologous but may be due to convergence. As with morphological systematic characters, meaningful comparisons can be made only between homologous proteins. The tertiary structure of a protein is an important characteristic in establishing homologies. All proteins which have so far been examined and judged homologous by their primary structures have similar tertiary structures. Homology of two proteins can be indicated by using various statistical methods which determine if the similarities are greater than could be due to chance. The situation is, however, complicated by the existence of gene duplication. The multiple globin and lactate dehydrogenase genes are, for example, homologous on the above definition.

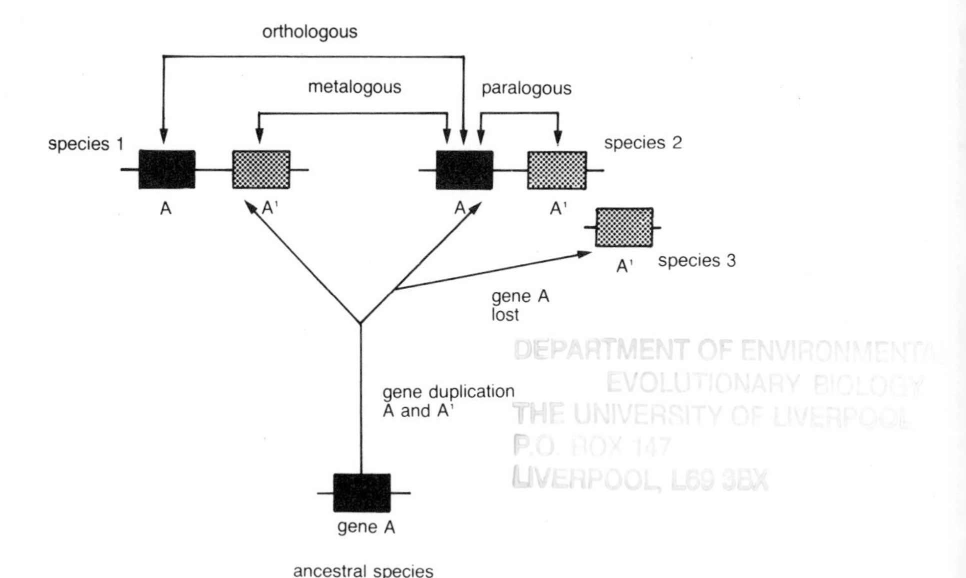

Figure 2.5. Types of homologous relationships between genes and their corresponding polypeptide products.

Figure 2.5 shows the different types of homologous relationships which can exist in situations where gene duplication has taken place. Equivalent genes or polypeptides in different taxa are said to have an *orthologous* relationship; those occurring within the same organism, a *paralogous* relationship. Non-equivalent duplicated genes in different taxa (e.g. the α and β haemoglobin genes) are called *metalogous*.

In cases such as the vertebrate globins, where the duplicate loci are retained, the choice for comparison is clear. However, if one duplicate gene is retained in one lineage and a different one in another lineage, then comparisons may be unknowingly metalogous and reveal genetic differences acquired since the original gene duplication, rather than since the two taxa diverged from a common ancestor (p. 131). The extent to which duplicate genes have been lost in polyploids suggests that this may present a serious problem. Two lysozymes are found in bird egg white: lysozyme c (originally described in the chicken) and lysozyme g (originally found in the goose). Both lysozymes c and g occur in some species of birds, but in general they are species-specific. The amino acid sequence of lysozyme c of the chicken is more similar to human lysozyme than it is to that of goose!

CHAPTER THREE

BIOCHEMICAL METHODS

SYSTEMATIC INFORMATION RELATIVELY FREE OF ENVIRONMENTALLY induced changes can be obtained by examining either the base sequence in DNA or the sequence of amino acids in proteins, the primary gene products. All other chemicals, structures and activities of the organisms are produced as a result of protein aggregation and enzymatic activity, or depend on the environment in which the organism is living. In plant systematics especially, many secondary chemicals of low molecular weight have been used. These substances include alkaloids, flavonoids, terpenoids, betalins, etc., and their use in plant systematics has been thoroughly reviewed by P. M. Smith (1976). Here we are concerned only with the systematic information that can be obtained from studies of the macromolecules, DNA and proteins.

While the classical systematist studies organisms by counting and measuring various parts and examining them under different types of microscopes, the biochemical systematist uses techniques borrowed from the protein chemist. Many detailed accounts of the theory and practice of these techniques are available (see Appendix, page 171). The following outline of the various techniques is intended only to provide sufficient background information for the reader to understand and evaluate results produced from such techniques. Emphasis will be placed on the limitations of the techniques and the need for adequate precautions and controls against artefacts.

Determination of primary structure of proteins

The maximum amount of systematic information is available when the complete amino acid sequence of proteins is known. Until recently, techniques for determining the primary structure of even quite small

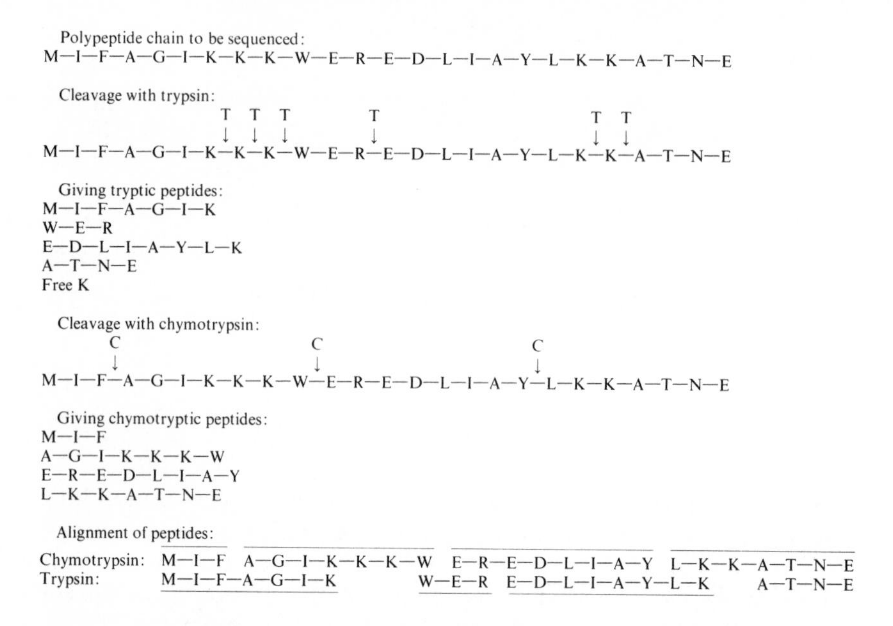

Figure 3.1. Determination of the sequence of a large polypeptide by comparison of the overlapping fragments produced by digestion with trypsin and chymotrypsin. See table 2.1 for single-letter amino-acid codes (after Wilson *et al.*, 1973).

proteins were too slow and labour-intensive to be of value in systematic work. Recently, however, the process has been automated, and sequences can now be determined relatively quickly.

Most procedures for determining amino-acid sequences make use of the Edman degradation process, whereby amino acids are sequentially removed from the amino-terminus of the polypeptide. In this procedure the reagent phenylisothiocyanate is coupled with the terminal amino group of the polypeptide to form a phenylthiocarbamyl derivative. Under anhydrous acidic conditions, the sulphur of the phenylthiocarbamyl group attacks the carbamyl component of the first peptide bond, resulting in cleavage of the terminal α amino acid as a thiazolinone. This cleaved amino acid is separated from the residual polypeptide by extraction with a solvent, and then converted to a more stable phenylthiohydantoin form prior to identification. The shortened peptide now has a new amino-terminal amino acid and can be subjected to further cycles of cleavage and identification of each successively removed amino acid, thus establishing the amino-acid sequence.

The cleaved amino acids can be identified using gas-liquid chromatography, thin-layer chromatography, or mass spectrometry or (after hydrolysis of the thiazolinones or thiohydantoins to free amino acids) by ion-exchange column chromatography in automatic amino-acid analysers. Unfortunately the Edman degradation can be carried out for only 60-70 cycles, and so large proteins need to be digested into small peptides for analysis. This digestion is carried out with two or more enzymes which break the polypeptide chain at different specific linkages. The endopeptidases trypsin and chymotrypsin are commonly used for this purpose. Trypsin breaks a polypeptide chain just after the point where the positively charged residues lysine and arginine occurs, while chymotrypsin similarly severs the chain after phenylalanine, tyrosine or tryptophan. The peptides resulting from digestion of a polypeptide with these two enzymes will be different but overlapping. From the sequences of these overlapping fragments, the overall sequence can be deduced (figure 3.1).

As well as the Edman degradation, which is not suitable for all proteins, cleavage with exopeptidases is used for primary-structure determination. Direct use of mass spectrometry has been employed in some cases.

The immunological approach

Proteins are antigenic, i.e. if they are introduced into an animal to which they are foreign, they are capable of causing the production of specific antibodies. Each protein has distinct sites on its surface called *antigenic determinants* against which antibodies are produced. The exact nature of these antigenic determinants and their relation to the amino-acid sequence of the protein is unknown. However, changes in the amino-acid sequence of a protein alter the nature of the antigenic determinants.

If a protein from another species is introduced into an animal such as a rabbit, specific antibodies to the antigenic determinants of that protein will be produced. These antibodies or immunoglobulins can be harvested from the rabbit as antiserum. If these antibodies are now mixed with the original protein in the presence of a suitable electrolyte, an antigen-antibody reaction will take place in which the combining sites of the antibody link onto the antigenic determinant sites of the antigen. If the antigen is multivalent (has several antigenic determinants per molecule), then since antibodies have at least two combining sites, large aggregates of antigen-antibody will form and will precipitate from solution if antigen and antiserum are present in approximately equal amounts.

If antibodies to a protein of one species are mixed with the same protein from another species (heterologous reaction), and if the two proteins have antigenic determinants in common, then an antigen-antibody reaction will take place. The fewer the determinants in common and the poorer the matching of the combining sites, the weaker the antigen-antibody reaction will be. Based on the magnitude of this, an immunological distance can be calculated. This is proportional to the difference in structure between a test protein and the protein used to prepare the heterologous antibodies.

In using this immunological approach for systematic purposes there are two main problems:

1. Obtaining suitable antibodies.
2. Accurately measuring the strength of the antigen-antibody reaction.

Preparation of antibodies

The rabbit is a convenient animal which is often used in the preparation of antibodies for systematic purposes. The main problem is that individual rabbits may respond differently to the same antigen.

If a protein is injected into several rabbits, each rabbit may produce antibodies to any or all of the antigenic determinants. The amount of antibody produced to any determinant may vary from rabbit to rabbit, and occasionally individual rabbits may fail to form antibodies to one or more determinants. The site of injection, the schedule and amounts of antigen used are also important in governing the antibody production. A prolonged immunization extends the range of antibodies produced and reduces the discriminatory power of the antiserum in heterologous reactions. Also the degree to which an antibody formed against protein A reacts with protein B may not be the same as in the reciprocal test where antibody formed against B is reacted with A.

Complement fixation

The extent of antigen-antibody reactions can be measured accurately by making use of the fact that a group of proteins found in serum, and known collectively as 'complement' binds onto antigen-antibody complexes. The greater the number of antigen-antibody reactions, the more complement is bound or fixed. In the complement fixation test, a known amount of complement is added to the antigen-antibody mixture, and the amount which is fixed is determined after a period of time. This is measured by using the property of free complement to lyse sensitized red blood cells (cells mixed with antibodies to the same type of cells), thus releasing haemoglobin, the amount of which can be measured in a spectrophotometer. The concentration of haemoglobin is inversely proportional to the amount of complement fixed into the antigen-antibody complex (figure 3.2). The micro method (micro-complement

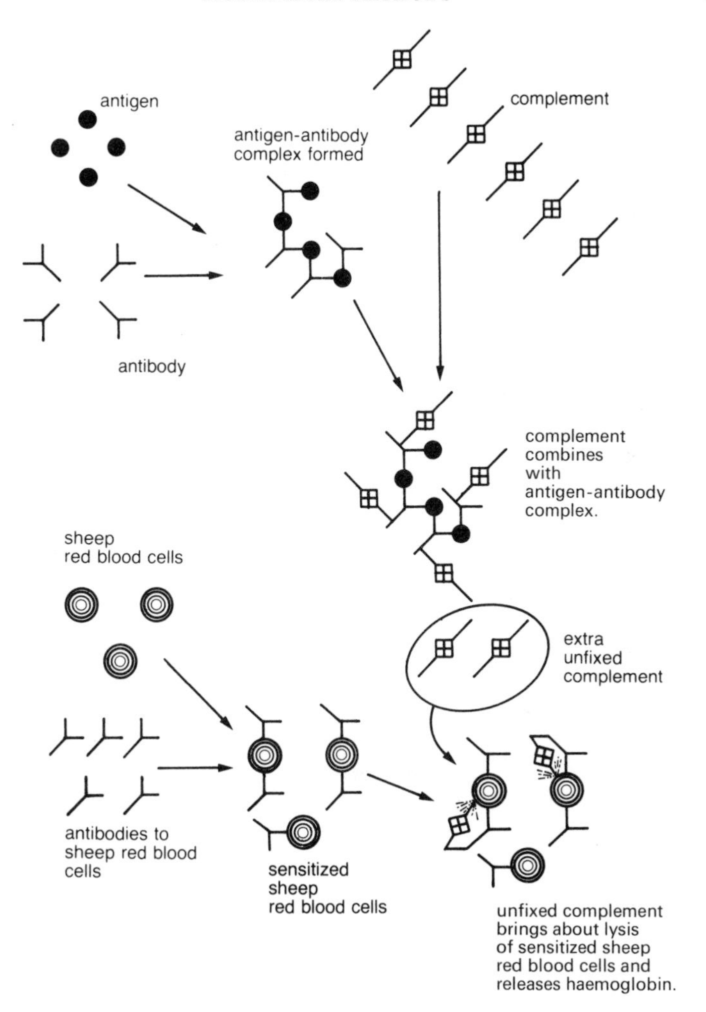

Figure 3.2. Steps involved in the complement fixation test.

fixation or MC'F) of Wasserman and Levine has been further developed by Wilson and his colleagues at the University of California and has been used widely for comparing amino-acid sequences among proteins (Champion *et al.*, 1974).

The amount of complement fixed varies with the concentration of the antiserum and is determined by titration. Since in the heterologous reaction the antigen-antibody reaction is weaker, a higher concentration of antiserum is needed to fix the same amount of complement. The factor by which the antiserum concentration needs to be raised for a particular antigen to produce the same complement fixation as the homologous antigen is an index of the dissimilarity. This can be used to calculate the immunological distance (ID) between the two antigens where ID = 100 × log (index of dissimilarity).

Immunological distance derived in this way has been shown to be linear with respect to amino-acid differences, suggesting that the antigenic effects of amino-acid substitutions are approximately equal and additive. Comparisons with proteins of known amino-acid sequences have shown that for bird lysozymes c and bacterial azurins the approximate relationship between immunological distance y and percentage differences in amino-acid sequences x is $y = 5x$, with a correlation coefficient of 0.9 (Champion *et al.*, 1975). For mammalian pancreatic ribonuclease the relationship is $y = 7x$ (Prager *et al.*, 1978). These relationships hold up to sequence differences of 30–40%. For the protein albumin which is a single polypeptide chain of some 580 amino acids, this means that one unit of immunological distance equals approximately one amino-acid substitution.

ELECTROPHORESIS

Of all the methods of biochemical systematics, the most widely used is *electrophoresis*. In general terms electrophoresis is the movement of charged particles under the influence of an electric field. All proteins carry an electric charge which is determined by their amino-acid composition and the pH of the medium (p. 18). In order to stabilize the charge, electrophoresis is carried out in a buffer or series of buffers. In an electric field, a protein will move towards the oppositely charged pole at a rate proportional to the magnitude of its charge. Since proteins may have different charges, they may move at varying rates and directions in electrophoresis. Electrophoretic mobility is used as an indicator of similarity of amino-acid composition when orthologous proteins are compared between individuals. The character examined is the mobility of a protein under a given set of electrophoretic conditions, and it exists in two states: same or different; the actual degree of mobility difference is difficult to interpret. Electrophoresis can be carried out in free solution but, for systematic work, some form of semi-solid supporting medium is used to enable detection of the relative positions of the proteins, after separation, as discrete bands or zones.

Sources and extraction of proteins

Proteins for electrophoresis must be in solution. Native protein solutions such as blood plasma, haemolymph, humoral fluids, milk, egg white, snake venom, etc., can be used directly or with dilution or concentration as required. Alternatively, aqueous or buffer extracts of tissue proteins can be made by homogenization, sonication, freezing and thawing or, simply, in the case of high-concentration proteins, by immersing the tissue for a short period of time. After disruption of the tissue, the mixture is centrifuged to remove insoluble debris which otherwise would clog the pores of the electrophoretic medium. The exact concentration of the buffer used is important, as changes in ionic strength and pH can bring about differential extraction of proteins. Membrane-bound proteins can be solubilized with detergents such as sodium dodecyl sulphate. In the case of small organisms, the entire individual or groups of individuals are used. A single unicellular animal such as *Amoeba proteus* can provide sufficient proteins for electrophoresis. For larger animals, proteins from most body tissues have been used, especially skeletal and heart muscles, liver and eye. In plants, leaves, seeds, flower buds and storage organs are used predominantly. Most proteins denature on removal from the *in vivo* state, and so studies are mainly confined to 'living' material or specimens stored under special conditions, normally frozen. Most of the preservative fluids used for museum storage of specimens denature proteins and make these specimens unsuitable for biochemical work. A few proteins, however, are extractable from fossils and from structures like hair and feathers of museum specimens. Suitable extracts have been prepared from fossil brachiopod shells (Jope, 1969). Proteins such as the collagens of bone and skin, and the keratins of various structures, are often well preserved through archaeological and even geological time (Jope, 1976). Seed and husk proteins also survive in archaeological deposits. Although tissue fluids, blood, etc., provide a useful source of proteins which can be obtained from live animals, to obtain a large number and variety of proteins involves extracting these from many different organs, and thus killing and dissecting of the specimen.

Once prepared, extracts can be stored frozen for periods of from days to years, depending on the proteins in question. Care is needed as storage may result in artefacts of electrophoretic patterns. Certain proteins, e.g. eye-lens proteins of birds, denature rapidly even when frozen (Sibley and Brush, 1967) and give aberrant electrophoretic patterns. Since in some tissues the presence of proteolytic enzymes can bring about protein digestion, extraction and all subsequent procedures should be carried out at as low a temperature and as quickly as possible. Low temperature also aids the retention of desired enzyme activity. In general, the lower the temperature of storage, the longer enzyme activity will be retained. Powers and Powers (1975) have noted that fish tissues stored at −190°C maintain 90% of their lactate dehydrogenase activity indefinitely. Storage at −40°C as opposed to −18°C results in the retention of activities of several enzymes for appreciably longer periods of time (author's own experience).

Plant material is especially difficult to work with as the disrupted tissues can release phenolic compounds which may bring about tanning of the proteins. In plants, where the protein concentration tends to be lower than in animal tissues, stable concentrated extracts can be produced by treating with cold acetone (−20°C) to yield dry powders which can be reconstituted in buffer when required for use.

Supporting media

The original support used in the early days of electrophoresis was filter paper, but now this has been replaced by gels of cellulose acetate, agar, starch or polyacrylamide. In supporting gels of starch or polyacrylamide, the pore size approximates to the molecular size of proteins, and large proteins are retarded more than smaller ones as they move through the gel. Thus a molecular sieving effect takes place and proteins, even if they have the same electric charge but are of different sizes or possibly shapes, may move at different rates during electrophoresis in these gels. Cellulose acetate and agar gels have large pores, and here separation is primarily on electric charge. However, large pore size means that there is little resistance to diffusion, and cellulose acetate and agar consequently are used for immunoelectrophoresis, i.e. electrophoretic separation of proteins followed by counter diffusion against specific antibodies (p. 40).

Starch gels are prepared by heating partially hydrolysed starch in a suitable buffer. Upon cooling, a semi-solid gel forms. Polyacrylamide gels are prepared by catalytic polymerization of acrylamide monomer and a cross-linking agent such as bis-acrylamide. The pore size of the gel formed is dependent on the concentrations of the acrylamide and bis-acrylamide. This ability to alter the pore size over a wide range is one of the most useful properties of acrylamide gels.

Electrophoresis in starch and polyacrylamide gels can be carried out in either horizontal or vertical modes. Large sample volumes cannot be applied to horizontal gels because of electrodecantation of the proteins to the bottom of the gel. This is overcome when the gel is placed in a vertical position.

Electrophoretic techniques

A multitude of variants on the basic electrophoretic technique have been developed over the past 20 years. The resolution of a particular technique is its ability to separate two similar proteins (i.e. for them to be of detectably different mobility). Resolution depends on the type of buffer employed and the supporting medium characteristics. As well as maintaining a constant pH in the separation medium, buffers are necessary to conduct the electric current. The two important features of a buffer are its pH and ionic strength. Ideally, the buffer used should have a pH as different as possible from the pI of the proteins to be separated in order to maximize the differences in their charges. Electrophoresis normally is carried out in the pH range 3–10 since outside this range irreversible denaturation of proteins takes place. Most electrophoresis is carried out in pH 8–9 buffers, at which pH most proteins are negatively charged and migrate towards the anodal end of the gel.

To separate two proteins it is necessary to permit electrophoretic migration to continue until one protein has travelled at least the thickness of the starting zone further than the other. However, the sharpness of the zones occupied by each protein diminishes with time because of the spreading of the zones as a result of diffusion. The thinner the starting zone in the direction of the electric field, the higher will be the resolution, until the point is reached where diffusion spreading of the zones' edges becomes large compared with the starting dimensions. The higher the potential gradient applied, the more rapid will be the electrophoretic separation, and hence the less diffusion which will take place. However, during the passage of the electric current through the supporting medium, heat is produced in proportion to the power (volts $\times$ amperes) applied to the gel. If too much power is applied, the excess heat will cause denaturation. This can be reduced by cooling the supporting medium, by placing it on a plate through which coolant is circulated, or by containing the entire apparatus within a refrigerator. The lower the ionic strength of the buffer, the lower will be the current through the gel, and hence the higher the voltage which can be applied without causing excess heating. On the other hand, buffers of low ionic strength have poorer buffering capacity.

With the combination of various supporting media, and buffers of different chemical make-up, pH and ionic strength, a bewildering range is presented to the investigator wishing to use electrophoretic techniques or interpret the results of others—they differ in their resolution, amount of protein required and ease of operation.

Electrophoretic techniques can be divided into five broad groups:

1. Continuous-buffer electrophoresis
2. Discontinuous-buffer or multiphasic electrophoresis and isotachophoresis
3. Isoelectric focusing
4. SDS and urea electrophoresis
5. Two-dimensional techniques

Continuous-buffer electrophoresis

In its simplest form, electrophoresis involves a supporting medium, a buffer to control pH and conduct the electric current, and a DC power supply (figure 3.3). In continuous-buffer electrophoresis, the buffer present in the tanks used to provide electric contact between the electrodes and supporting medium is the same as in the medium itself. A continuous-buffer system is used normally in cellulose acetate, agar electrophoresis and also in acrylamide gels where the separation is on

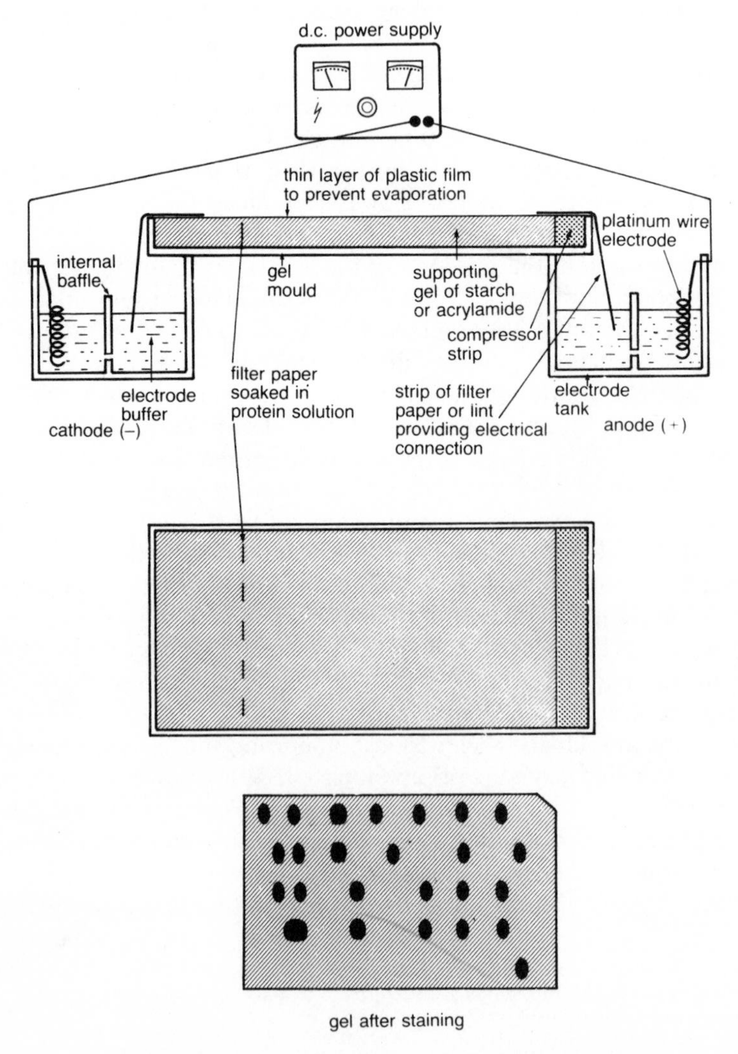

Figure 3.3. Apparatus for horizontal gel electrophoresis.

molecular size alone. In this latter technique, gels are prepared with a continuously varying pore size from one end of the gel to the other. If a mixture of proteins is applied to a gradient gel at the large pore end and an electric potential applied across the gel, the proteins will migrate until they reach the pore size in the gel at which they can move no further. Individual proteins will become fixed at their respective pore-limits. In this *gradient-pore* technique, the electric charge is used only to move the proteins to their pore limit, and therefore separation is on the basis of molecular size.

Multiphasic electrophoresis

In continuous-buffer electrophoresis, to achieve maximum resolution the sample has to be applied in a very narrow starting zone (with the exception of the gradient-pore technique). In the early work on electrophoresis in starch gels, it was found that if a different buffer was used in the electrode tanks from that in the gel, considerably improved resolution resulted. Discontinuous-buffer systems have been exploited to a large extent in polyacrylamide gels where they can be combined with regions of different pore sizes. In these multiphasic systems, the first part of the gel has a large pore size which exerts little sieving effect, and provides a region where buffer discontinuities are used to concentrate the individual protein components in the sample into narrow stacks. This means that the proteins enter the second part of the gel as very narrow zones, and it is in this second small-pore region that electrophoretic separation takes place on the basis of electric charge and conformation. This technique is sometimes known as *disc electrophoresis* because the proteins stack as a series of concentrated discs in the first part of the gel. This stacking involves the 'sandwiching' of the proteins between a buffer ion of faster mobility than any of the proteins (leading ion) and an ion of slower mobility (trailing ion). When an electric potential is applied, the proteins arrange themselves in stacks in order of their mobilities between the leading and trailing ions. This stacking phase is known as *isotachophoresis* and can be used as an analytical tool on its own.

Isoelectric focusing

Isoelectric focusing is an electrophoretic technique using large-pore polyacrylamide gel in which is incorporated a mixture of synthetic polyamino polycarboxylic acids ('carrier ampholytes') with a range of isoelectric points. When an electric potential is applied to the gel, the ampholytes form a stable pH gradient from one end of the gel to the other. The ampholytes are confined to the gel by using a strong acid at the anode and a strong base at the cathode.

When a mixture of proteins is introduced into this pH gradient, the various proteins will move electrophoretically until they reach the point on the gel where the pH is equal to their isoelectric point (pI). At this point the protein is electrically neutral and will move no further. Should it diffuse from this point, it will develop charge and move back to the pI point to become concentrated or focused into a very narrow zone on the gel—hence the name, isoelectric focusing. This method gives very high resolution and proteins differing by 0.01 unit in their pI values can be separated.

SDS and urea electrophoresis

Urea and the anionic detergent sodium dodecyl sulphate (SDS) are capable of solubilizing certain classes of proteins and also of breaking polymeric molecules into constituent polypeptides. SDS binds to polypeptides and imparts a large negative charge which masks the individual variation in electric charge. In SDS-containing gels, polypeptide migration is dependent solely on molecular weight, and with the use of suitable known markers can be used to give an estimate of this parameter. In urea gels the normal charge-conformation separation takes place

Two-dimensional electrophoresis

Any pair of the above electrophoretic techniques can be used in sequence to give a two-dimensional separation, i.e. separation of a protein mixture by one technique followed by separation at right angles to the original direction of movement by another. The most commonly used combination is isoelectric focusing in the first dimension and SDS-electrophoresis in the second. This results in separations on the basis of two independent parameters, isoelectric point and molecular weight.

Detection of proteins after electrophoresis

Since most proteins, with the exception of haemoglobin and a few others, are colourless, the supporting medium must be stained to reveal the positions of the proteins at the end of the electrophoretic run. Staining can be carried out, either by using a non-specific dye such as Amido Black or Coomassie Brilliant Blue R, which stain all proteins present in sufficient concentration, or by more specific stains for glycoproteins, lipoproteins, etc. In the case of enzymes, the particular enzyme activity can be used to locate their position after electrophoresis. In most cases the enzyme is used to break down the specific substrate, and in so doing to bring about the oxidation or reduction of a soluble chemical to an insoluble coloured form. Thus a coloured precipitate forms at the site of the enzyme.

Esterases can be detected due to their hydrolytic activity on the artificial substrate α-naphthyl acetate producing α-naphthol which combines with tetrazotized o-dianisidine (Fast Blue B salt) or other diazotized aromatic amine, to give a coloured precipitate. Any enzyme which brings about the reduction of either of the pyridine nucleotides, nicotinamide-adenine dinucleotide (NAD) or nicotinamide-adenine dinucleotide phosphate (NADP) can be stained by the tetrazolium method. This involves the reduction of a colourless soluble tetrazolium salt (e.g. nitro-blue tetrazolium, NBT) to an insoluble blue formazan. The staining mixture also contains an electron carrier, phenazine methosulphate (PMS). The reactions involved for the detection of lactate dehydrogenase are shown in figure 3.4.

In cases where the enzyme of interest does not bring about a reduction of NAD/NADP, other enzymes which react with the product of the first reaction to produce this change may be added to the staining mixture. For example, in the detection of phosphoglucose isomerase (PGI) this enzyme reacts with the substrate, fructose-6-phosphate, converting it to glucose-6-phosphate. The staining mixture contains glucose-6-phosphate dehydrogenase which reacts with the glucose-6-phosphate produced, resulting in the dehydrogenase reaction as above. Methods for other enzymes can involve two or three coupling enzymes before a detectable reaction occurs.

Most laboratory chemicals are less than 100% pure and contaminants can result in reactions other than the desired one taking place, thus producing additional bands on the gel, e.g. gels on which liver samples

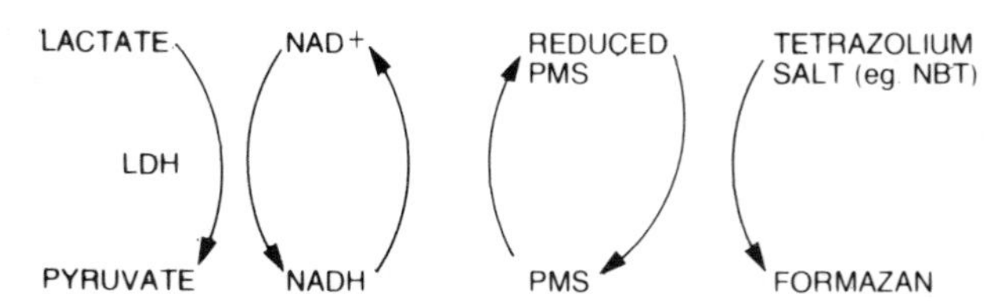

Figure 3.4. Detection of lactate dehydrogenase by the tetrazolium method.
LDH = lactate dehydrogenase
NAD^+ = nicotinamide adenine dinucleotide
NADH = reduced nicotinamide adenine dinucleotide
PMS = phenazine methosulphate
NBT = nitro-blue tetrazolium

have been run often produce dehydrogenase bands if incubated with NAD, PMS and NBT, but without substrate. This 'nothing dehydrogenase' has been shown to be alcohol dehydrogenase, sufficient alcohol being present as a contaminant to act as substrate.

If, as with glucose-6-phosphate, the intermediate reaction product is soluble or if the final coloured product is soluble, then the staining reagents can be mixed with molten agar which sets on the surface of the gel as an overlay. This confines the reaction products to the site of the enzyme activity. Alternatively, the staining solution can be absorbed by filter paper or cellulose acetate which is then laid on the surface of the gel. The pattern of bands which results after staining a gel for enzymes is called a *zymogram*, that from general protein staining an *electrophoregram*.

Immunological localization and identification of proteins

If, after electrophoresis, a trough is cut in the gel parallel to the direction of migration, and this trough is filled with antiserum containing a mixture of antibodies raised against the proteins of the sample, the proteins (antigens) will diffuse through the gel and form precipitin lines where they meet with complementary antibodies. Homologous and heterologous reactions can be compared, and the number of common antigens estimated. The combination of electrophoresis and immunodiffusion is called *immunoelectrophoresis* and is normally carried out in agar gels, as this is the best medium for diffusion and antigen-antibody precipitation. The electrophoretic step can be carried out in starch or acrylamide, followed by embedding of a strip of this gel in agar for the diffusion stage to take place.

Instead of allowing the antigen and antibody molecules to diffuse together, they can be brought together electrophoretically. A strip of gel in which the first electrophoretic separation has been carried out is placed at the edge of an agar gel containing antibodies, and electrophoresis is carried out at right angles to the first separation. If this electrophoresis is performed at pH 8.6, then the gamma globulin antibodies have little or no mobility, but the proteins will move through the antibody-containing agar gel. This is much quicker than diffusion, and results in rocket-shaped lines of precipitation. This technique is known as *Laurell* or *crossed immunoelectrophoresis*. A modification of it is *tandem crossed immunoelectrophoresis* where two mixtures of proteins which are being compared are put on the electrophoresis gel a few millimetres apart. After completion of the second-dimension electrophoresis, two almost superimposed patterns result, enabling side-by-side comparison of the homologous and heterologous reaction.

Comparison of electrophoretic techniques

Ideally, the electrophoretic technique used should be capable of resolving all the proteins present in an extract, and the pattern should be completely reproducible. A comparison of the relative resolving powers of various electrophoretic techniques can be made by reference to separation of vertebrate plasma which contains some 100–200 proteins. Electrophoresis of plasma on cellulose acetate reveals 5 protein bands, on starch gel 15 bands, acrylamide with a discontinuous buffer 19 bands, and isoelectric focusing in the pH range 3.5–10 gives 30+ bands. These figures indicate only roughly the resolving power and vary with the precise conditions used. With two-dimensional isoelectric focusing in urea/SDS-electrophoresis, human plasma resolves into 300 or so protein spots which, due to carbohydrate and other heterogeneity, represent perhaps 75-100 polypeptides (Anderson and Anderson, 1977). Similar two-dimensional separation of extracts from the bacterium *Escherichia coli* gives 1100 protein components (O'Farrell, 1975).

Electrophoresis can either be carried out on individual gels or on slab gels which permit a number of samples to be run together. Since systematics is a comparative study and involves comparisons of mobilities of homologous proteins in different samples, this side-by-side separation where all samples will have been subjected to as near as possible identical conditions is desirable. Minor variations in pH, ionic strength, purity of chemicals, gelling temperature, running temperature, electric field, etc., can alter protein mobilities. Thus inter-gel comparisons can be unreliable unless considerable care is taken in the standardization of techniques. Even then, a purified protein such as haemoglobin or ferritin (Johnson, 1975) or bovine serum albumin should be used as standard control on each gel. The test proteins' mobilities can be related to the mobility of the standard. Such relative mobility values are amenable to analysis by numerical taxonomic methods.

If the gel is stained with a non-specific general protein stain, each band may represent two or more proteins, since no electrophoretic technique resolves all the proteins present in most tissue extracts. A further difficulty in the use of general protein patterns is the problem of homology. Bands of equivalent position on the gel in samples from different species may not represent orthologous proteins. Also, variable proteins may be superimposed and conceal each other's phenotypic variation. For these reasons it is very difficult to evaluate the genetic basis of intra-specific variability from general protein patterns. For most systematic work, staining for specific enzymes is preferable, and reduces the problems of homology and interpretation of variability.

In staining for specific enzymes, a maximum resolution technique may not be required, e.g. although starch gel gives poorer resolution than acrylamide for general proteins, enzyme staining is in most cases superior. The apparatus for starch gel is also simpler and has the further advantage that a standard 6 mm gel can be divided into a number of 1-mm slices, and each stained for a different enzyme. Acrylamide gels have the disadvantage that the acrylamide monomer and bis-acrylamide are neurotoxic, and there is a serious risk of poisoning by inhalation or skin contact. Even in polymerized gels some monomer remains. For certain problems, high-resolution techniques such as isoelectric focusing are valuable. Isoelectric focusing has the disadvantage that the ampholytes are expensive. Where small organisms or a dilute sample have to be used, the concentrating stage of multiphasic acrylamide electrophoresis can be employed to advantage.

Limitations and artefacts of electrophoretic techniques

Since electrophoresis is the major technique of biochemical systematics, it is appropriate to consider its limitations and possible sources of error which might arise from uncritical use. Electrophoresis examines the overall electric charge and/or conformation of a protein. Sixteen of the common amino acids have non-ionizable side-chains and are electrically neutral in the pH-range employed in electrophoresis. Proteins possess charge as a result of glutamic and aspartic acids (negative) and lysine and arginine (positive) components (p. 18). Substitution of an amino acid for a like-charged one will have little or no effect on the overall charge, and an insignificant effect on the molecular weight. What then is the probability that an amino-acid substitution will result in a change in charge? Examination of the genetic code shows that of the 399 possible non-redundant single base changes, 128 or 32% will result in the substitution of an amino acid of different charge (Lewontin, 1974). These values take no account, however, of the inequalities in amino-acid content of proteins. Examination of proteins of known sequences gives a figure of 27.5%. Whenever proteins differ by two or more substitutions involving charged amino acids, then there is the possibility of substitutions of two oppositely charged amino acids cancelling each other, e.g. the cytochromes c of organisms which differ greatly in their amino-acid sequences (p. 135) all have compensatory substitutions and are identical in the arithmetical sums of their charges (Lewontin, 1974). Application of very sensitive electrophoretic techniques can, however, detect the conformational differences. This then is the crux of the electrophoretic method:

differences can be detected, but not similarites.

The fraction of amino-acid substitutions which can be detected by electrophoresis can be increased by sequential use of different techniques. Thus electrophoresis is based on three principal properties of proteins: net charge, isoelectric point, and size and conformation. Some proteins may have the same net charge at one pH and not at another. If two proteins have the same mobility in electrophoresis using buffers of different pH, in isoelectric focusing, and in porosity gradient or SDS electrophoresis, then a much higher probability can be attached to their being the same in amino-acid sequence than if they were examined by a single type of electrophoresis. In addition, catalytic properties and resistance to thermal and other types of denaturation can be used to test further for differences.

Apparent differences in electrophoretic mobility may be due to artefacts in the gel, or to some form of post-translational modification of the proteins. These latter changes may result from denaturation, deamination, phosphorylation, sulphation, oxidation, reduction, addition of other molecules, aggregation, and cleavage of polypeptides. The following examples will show some of the problems which result from these changes. Homozygous transferrin of pigeons (*Columba livia*) exhibits three bands after electrophoresis on starch or acrylamide. This heterogeneity has been shown to be due to different amounts of the carbohydrate sialic acid attached to the transferrin, and digestion of the sample with neuraminidase prior to electrophoresis results in a single band. Similar sialic acid-induced heterogeneity has been shown for alkaline phosphatase. Binding of iron to ovotransferrin results in an increase of negative charge and increased electrophoretic mobility (Baker *et al.*, 1966; Stratil, 1967). In borate buffers, albumin gives two bands due to attachment of borate ions to the protein molecule (Cann, 1966). Similarly, attachment of ampholytes to various proteins can result in multiple bands on isoelectric focusing (Hare *et al.*, 1978). Residual ammonium persulphate in acrylamide gels can cause splitting and disappearance of bands (Mitchell, 1967; Brewer, 1967; Fantes and Furminger, 1967). Binding of NAD or NADP can change the mobility of proteins. Thus NAD increases the mobility of lactate dehydrogenase (Wilkinson, 1970) while NADP can change the electrophoretic behaviour of glucose-6-phosphate dehydrogenase (Bakay *et al.*, 1972) as can the anticoagulant heparin which is frequently used in the collection of plasma samples (Bergman *et al.*, 1971).

In a comparative study, if all samples are collected, stored and analysed in an identical manner, then most of the above problems probably do not arise. However, for systematic work it is often necessary to store samples for lengthy periods until the complete set of specimens is acquired. Appropriate controls are necessary to ensure that no artefacts are introduced in this way.

Contamination by exogenous proteins

When an extract of a tissue or whole organism is prepared, proteins from other than the organism concerned may be included. Two main sources are parasites and food present in the gut. Additional esterase bands were found in zymograms of eye lenses of the fish *Poeciliopsis lucida* which were infected by metacercariae of the trematode parasite *Neascus* sp. (Vrijenhoek, 1975*a*). Allen and Gibson (1975) have shown that enzyme bands contributed by bacteria can result in confused electrophoretic patterns for *Paramecium*, and they point out the need to use axenic (bacteria-free pure cultures) of protozoans. Again enzymes and other proteins derived from food are present in the gut. Indeed electrophoresis of these enzymes has been used to determine which species of prey had been eaten by particular predators (Murray and Solomon, 1978).

Peptide analysis—'finger-printing'

Proteins can be broken up into small chains of amino acids by digesting with enzymes which break specific amino-acid linkages (p. 29). The mixture of peptides obtained in this way can be separated into individual peptides by two-dimensional electrophoresis-chromatography. The mixture of peptides is placed at one corner of a sheet of filter paper or thin layer gel of silica, cellulose, etc., and high-voltage electrophoresis is carried out. This moves the peptides along one edge of the support. Subsequently chromatography is carried out at right angles to the direction of electrophoresis. Since electrophoresis and chromatography act on different properties of the peptides, they respond differently as they move among the L-shaped separation path, and so individual peptides separate one from another. The position of the peptides can be located by a suitable chromagenic reaction. The resulting pattern of spots or 'finger-print' can be compared for orthologous proteins in various taxa. Ion-exchange column chromatography followed by elution through a suitable detector system also can be used to analyse peptide composition of a protein.

Two proteins which appear similar on electrophoresis can be eluted from the gel and subjected to peptide analysis. There is a very high probability that the two proteins are of the same primary structure if they appear identical using this method.

NUCLEIC ACID ANALYSIS

The techniques for the determination of base sequences of nucleic acids are still in their infancy but have developed considerably over the past few years. These methods involve selective cleavage of nucleic acid by enzymes (Simoncsits *et al.*, 1977) or chemical reactions (Maxam and Gilbert, 1977), or copying a section of the molecule (Sanger and Coulson, 1975). Most nucleic acid sequencing has been restricted to micro-organisms and as yet has not been applied to systematic problems. This approach undoubtedly will be important in the future. However, even if a technique could be devised to sequence DNA at the rapid rate of one base per second, it would still take four months to determine the complete base sequence of a bacterium and over 100 years for a mammal (Powell, 1975). For the forseeable future, then, there will be a requirement for short-cut methods such as electrophoresis.

Restriction endonuclease analysis

Double-stranded DNA can be cleaved into short sections by the use of enzymes which recognize specific nucleotide sequences. These cleavage site-specific endonucleases are analogous to specific proteolytic enzymes, and are proving as useful in the study of the structure of DNA as trypsin and chymotrypsin have been in protein sequencing. The DNA fragments produced by these enzymes can be separated on electrophoretic gels on the basis of their size. Potter *et al.* (1975) compared the banding patterns shown on electrophoretic gels after separation of the products of restriction endonuclease analysis of mammalian mitochondrial DNA. Phylogenetically closely related species showed a greater number of DNA bands of equivalent mobility than species which are less closely related.

DNA hybridization

An indirect approach to studying DNA base sequence is by the method of DNA hybridization. Separation of the two strands which make up the DNA double helix can be brought about *in vitro* by heating. When allowed to cool again, the double molecule or duplex will be reformed due to the specific base-pairing requirements. However, because of some errors in pairing between some of the bases, a molecule of lower overall bonding strength will result. If this re-annealed DNA is heated again, it is found that, as a result of these errors, a lower temperature is required to separate the two strands as compared with the original native DNA. The DNA from different species can be brought together to form hybrid duplexes. The more similar the two species are in their DNA base sequence, the greater will be the pair matching in this hybrid duplex, and the higher the thermal stability. The difference between the temperature at which reformed DNA of one species dissociates and that of the hybrid is proportional to the similarity of the base sequence of the two species. The relationship between thermal stability and degree of mismatching is linear, and such that about 1.0% base pair mismatching will lower the dissociating temperature by 1°C at 50% dissociation. Thus the lower the temperature necessary to melt the hybrid duplex, the greater the genetic differences between the two species.

The most useful information comes from hybridization experiments with single-copy DNA sequences (p. 143). The highly repetitive DNA sequences are readily detectable and removed by their rapid rate of reassociation relative to the rate of reassociation of unique DNA. Two complementary strands must collide in order to reassociate and, the more frequent a sequence, the more collisions will take place. A sequence which is repeated 1000 times will reassociate 1000 times more rapidly than one which is present once.

DNA sequences differing by more than about 20% are unable to form stable duplexes, and so this method cannot be used to compare distantly related species.

CHAPTER FOUR

INTRA-SPECIFIC VARIATION

BEFORE THE DIVERSITY AMONG SPECIES AND HIGHER TAXA CAN BE considered, the variation which is present within species must be taken into account. Since comparisons at this level involve the examination of a large number of specimens, electrophoresis is the main technique employed, as it is sufficiently rapid in use and also sensitive enough to detect minor differences. Two individuals which belong to the same population or species may differ from each other for a number of reasons (p. 10). The main factors which influence electrophoretic studies of proteins are genetic variation, age, sex, physiological condition, and environment. Many changes which occur in electrophoretic patterns under the influence of these conditions (except for genetic variation) are quantitative, i.e. changes in amount of a particular protein rather than in its mobility. For the most part, only qualitative (mobility) differences are of value in systematics. However, after allowing for non-genetic changes, studies of quantitative variation in orthologous proteins among taxa may be of value in determining regulatory gene differences (chapter 9).

Isozymes

In many organisms, within the same individual there are enzymes with similar catalytic properties but differing in primary or higher structure. These enzyme forms are called *isozymes* (or isoenzymes) and are the products of separate genes. Isozymes can be formed by post-translational modification, possibly as a result of inter-gene (epigenetic) influences. Enzymes which exist in different configurational states but which have the same primary structure are called *conformers*. Malate dehydrogenase often shows a series of bands of different electrophoretic mobility which appear to be due to different stable configurations and not caused by any difference in primary structure.

Isozymes appear to be necessary to catalyse the same reaction under different metabolic conditions and frequently they are therefore differentially expressed at stages in the life history, in different organs or in different parts of the same cell, e.g. mitochondria and cytosol. Since isozymes have different structures they may be detected electrophoretically as bands of different mobility which react with the same substrate. Subsequent to 1959, when the existence of isozymes was first recognized by Markert and Møller, many enzymes in most organisms have been shown to exist in isozymic forms.

In most vertebrates there are three genes coding for the enzyme lactate dehydrogenase:

A which is predominantly active in skeletal muscle;
B in heart muscle;
C in eye retina, liver (also testis of birds and mammals).

In tissues of some species both A and B genes are simultaneously expressed. Lactate dehydrogenase is a tetramer and, since in many cases A and B type polypeptides can combine, five types of tetramer are produced: two homotetramers (A^4 and B^4) and three heterotetramers (A^3B^1, A^2B^2, A^1B^3). This results in five isozymes and a five-banded lactate dehydrogenase zymogram for these tissues. In some species of fish, only the homotetramers (and sometimes also the symmetrical heterotetramer) are revealed, due presumably to restrictions on subunit assembly or differential stability (Shaklee, 1975).

From the systematist's point of view, the existence of isozymes means that care must be taken to examine homologous life stages, tissues and parts of cells when making inter-taxon comparisons. Fish, for example, have a number of enzymes with different isozymes in the 'red' lateral line muscle from those in the 'white' muscle, and considerable care is required when taking samples. Contamination of a tissue with blood can give a similar effect.

Age-dependent changes

As noted already, isozymes may be differentially expressed during the life cycle. The same applies to non-enzymatic proteins. Human adult haemoglobin is principally HbA consisting of two α and two β polypeptides, whereas in the foetus more than half is HbF which is composed of two α and two γ chains. Haemoglobin F disappears completely during infancy. In very young embryos (up to approximately the third month) a fourth type of polypeptide, ε, is present and this

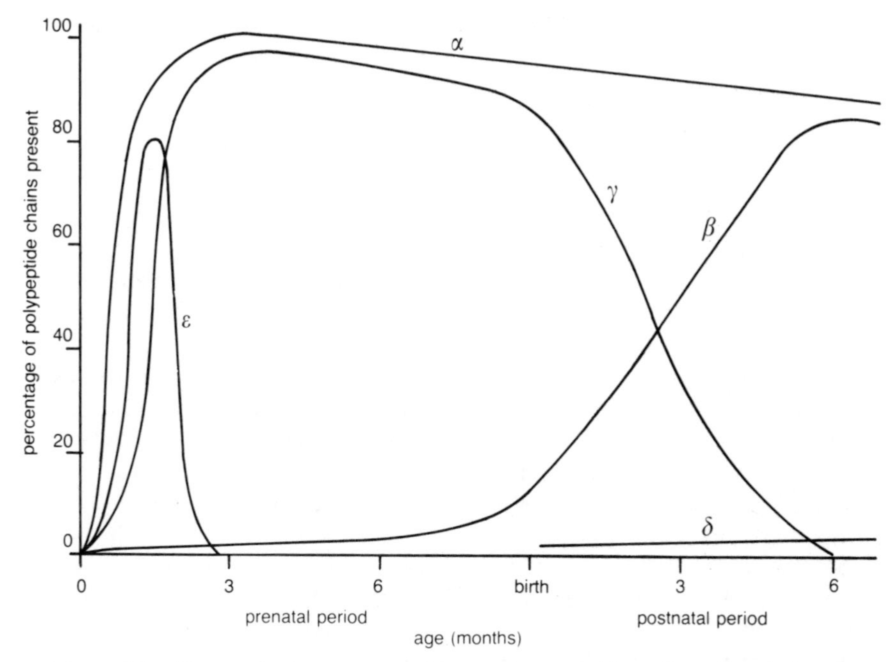

Figure 4.1. Changes in the amounts of human haemoglobin polypeptides during the prenatal and early postnatal periods (redrawn from Stansfield, 1977).

produces two embryonic haemoglobins: ε_4 (Hb Gower 1) and $\alpha_2\varepsilon_2$ (Hb Gower 11) (figure 4.1).

Particular proteins may be produced only at certain stages in the life history. In some chironomid midge species, haemoglobin is present in the larvae but is absent from the adults. There are ontogenetically linked changes also in the number of haemoglobin types found in the four instar stages.

Sex-associated changes

Occasionally proteins which are limited to one sex have been found, e.g. a kidney esterase in mice (Shaw and Koen, 1963), an amylase in *Drosophila hydei* (Doane *et al.*, 1975) and lactate dehydrogenase C in testis of birds and mammals. Undoubtedly there are a number of sex-limited proteins restricted to the gonads. Sex-specific proteins may be of value in determining the sex of an individual before the gonads are morphologically developed.

Several proteins are sex-linked, i.e. their genes are on the X chromosome. Examples are the enzymes, glucose-6-phosphate dehydrogenase and phosphoglycerate kinase in man and kangroo (Cooper *et al.*, 1975) and esterase-5 in *Drosophila pseudoobscura*. Due to inactivation of one of the X chromosomes in the homogametic sex, heterozygote expression is absent.

Environmental and physiological changes

Environmental factors and disease may result in differential expression of isozymes and changes in other proteins. In the killifish *Fundulus heteroclitus* kept in the dark, one lactate dehydrogenase isozyme disappears and a different one is expressed (Massaro and Booke, 1971). Qualitative differences in isozyme expression have been noted in yeasts grown under aerobic or anaerobic conditions. Certain breeds of sheep when made anaemic produce a new type of haemoglobin. In the rainbow trout *Salmo gairdneri*, both qualitative and quantitative changes have been found among individuals acclimated at different temperatures. This appears, however, to be an exception rather than a general rule in fish (Shaklee *et al.*, 1977). In mammals there is a close relationship between the expression, in the liver, of the five lactate dehydrogenase isozymes and dietary habits (Ogihara, 1975) which is shown as both an intra- and an inter-specific effect. Changes in isozymes associated with disease states have been well documented and are valuable as diagnostic aids (Vessel, 1975).

Polymorphism

Most higher organisms are diploid, and sexually reproducing forms receive one complete set of chromosomes from each parent. The gene (or *locus* as it is more specifically called) which directs the synthesis of a polypeptide is composed of two parts or alleles. An *allele* is the corresponding base sequence on each member of the homologous chromosome pair, and hence one allele comes from each parent. The two alleles code independently for the same polypeptide, i.e. each directs the production of half of the amount of that polypeptide. However, if one of the alleles contains a different codon, then that locus will produce two polypeptides differing from each other by an amino-acid substitution. When, *within a species*, the most common allele at a locus is of a frequency of less than 0.99 (some definitions use 0.95) then that locus is said to be *polymorphic*. Rare alleles at lower than this frequency do not come within the definition of polymorphism. Since slightly different forms of the protein coded for by a structural locus may be produced by these alternative alleles, this protein is said to be polymorphic also, i.e. it exists in a number of morphs or forms.

Within the species as a whole, a large number of alleles, differing from each other mainly in single bases, may exist. In the butterflies *Colias eurytheme* and *C. philodice* there are 25–30 alleles at an esterase locus in single populations (Burns, 1975). In man there are some 200 variants of adult haemoglobin as a result of alternative alleles at the α and β loci.

In each individual of a species two conditions are possible: either the alleles at a particular locus are identical (homozygous condition) or different (heterozygous condition), e.g., if, in a population, two alleles exist for a particular locus and we denote these alleles by A and B, then there are three possible genotypes: AA, AB and BB. In each of the homozygotes AA and BB one type of polypeptide will be produced, but in the heterozygote two types will result as, at the level of structural gene expression, almost all alleles are co-dominant, i.e. both are equally expressed. Because of the redundancy of the genetic code not all alleles will result in the production of different polypeptides. The lack of dominance at the protein phenotype level makes the genetics of protein polymorphism very straightforward. Since in sexually reproducing organisms one of the alleles at a locus comes from each parent, all the offspring of an individual homozygous for allele A mating with an individual homozygous for allele B will be AB heterozygotes.

If the proteins coded for by alleles A and B differ by a single amino acid, and this substitution results in a change in electric charge or in conformation of the molecule, then they will have different electrophoretic mobilities. So within the limitations of the technique, electrophoresis provides a convenient method of determining genetic variation at structural loci. Clearly, to estimate such variation, a large number of individuals needs to be examined, and this eliminates the use of techniques such as nucleic acid and protein sequencing.

In the example above, if, from a number of individuals, the tissues where this protein is expressed are sampled, extracted and subjected to electrophoresis, then after appropriate staining of the gel, three electrophoretic patterns will be shown (figure 4.2*a*, 4.3). The situation can be described quantitatively by stating the relative proportions of the three genotypes and the two alleles, i.e. the genotypic and allelic frequencies (p. 162). In the absence of dominance (and within the limitations of the electrophoretic method), phenotype is equivalent to genotype, and the numbers of each genotype are obtained simply by counting the number of individuals with each of the three patterns.

If three alleles A, B and C exist within a species, and each results in a protein with electrophoretically distinct mobilities, then six electrophoretic patterns will be found (figure 4.2*b*). In the heterozygote, normally only two bands are found on the gel if the protein is a monomer (i.e. it consists of a single polypeptide subunit). For a dimeric protein, due to the production of two different polypeptides in the

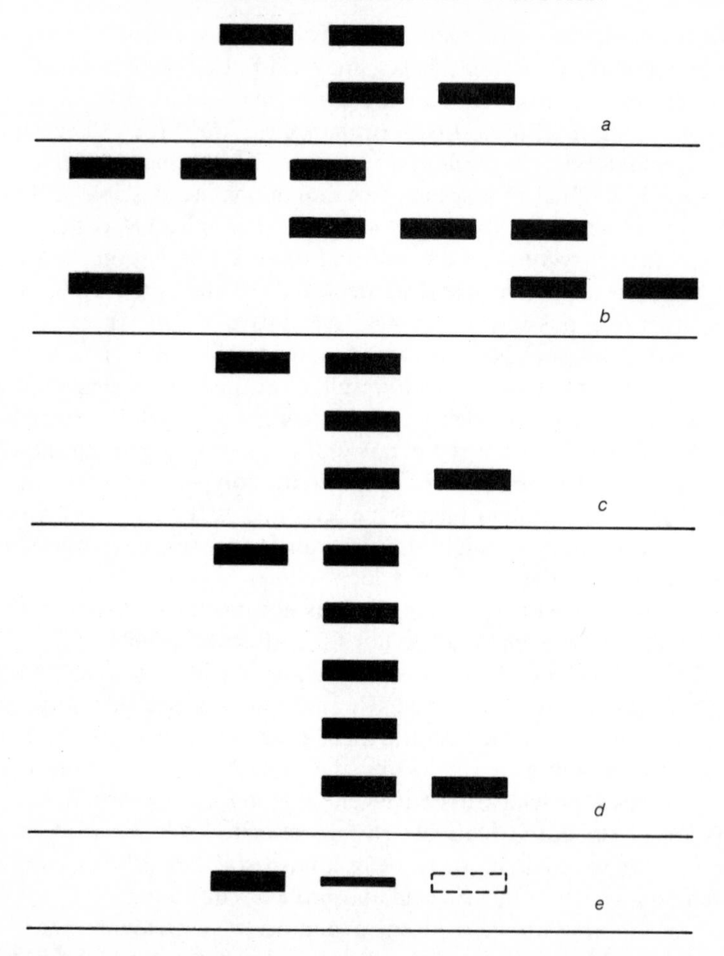

Figure 4.2. Patterns of protein polymorphism: (*a*) two codominant alleles—monomeric protein; (*b*) three codominant alleles—monomeric protein; (*c*) two codominant alleles—dimeric protein; (*d*) two codominant alleles—terrameric protein; (*e*) one normal and one null allele.

heterozygote, three different dimers are found (aa, ab, bb), i.e. there is a hybrid dimer produced (figure 4.2*c*; 4.4). About half of the proteins commonly screened in electrophoresis are dimers (table 4.1). For tetrameric proteins five different tetramers consisting of two polypeptides are possible (a_4, b_4, a_3b_1, a_2b_2, a_1b_3) resulting in a five-banded heterozygote pattern (figure 4.2*d*; 4.5). In some proteins there is restriction on subunit assembly, and hybrid molecules are not found. There is variation from taxon to taxon in this respect, even among homologous proteins.

Table 4.1 Specific enzymes and other proteins commonly examined by electrophoresis in systematic studies

Enzyme	*Abbreviation and locus*	*Subunit number**
Alcohol dehydrogenase	ADH_1	2
	ADH_2	2
	ADH_3	2
Glycerol-3-phosphate dehydrogenase	GPD_1	2
(α glycerophosphate dehydrogenase)	GPD_2	2
Sorbitol dehydrogenase	SORDH	4
Lactate dehydrogenase	LDH_A	4
	LDH_B	4
	LDH_C	4
Malate dehydrogenase	$MDH_{cytosol}$	2
(NAD–MDH)	MDH_{mito}	2
Malic enzyme	$ME_{cytosol}$	4
(NADP–MDH)	ME_{mito}	4
Isocitrate dehydrogenase	$ICD_{cytosol}$	2
Phosphogluconate dehydrogenase	PGD	2
Glucose-6-phosphate dehydrogenase	Gd	2
Superoxide dismutase	SOD_A	2
(tetrazolium or indophenol oxidase)	SOD_B	4
Glutamate-oxaloacetate transaminase	$GOT_{cytosol}$	2
(aspartate aminotransferase, AAT)	GOT_{mito}	2
Creatine kinase	CK_{muscle}	2
	CK_{brain}	2
Phosphoglucomutase	PGM_1	1
	PGM_2	1
	PGM_3	1
Esterase	ES (several)	1/2
Peptidase	PEP_A	1
	PEP_B	1
	PEP_C	1
	PEP_D	2
Enolase	ENO	2
Glucose phosphate isomerase	GPI	2
(phosphoglucose isomerase, PGI)		
Haemoglobin	Hb	4
Transferrin	Tf	1

*In man, but similar tertiary structure in most vertebrates. Enzyme data from Hopkinson *et al.* (1976).

A further condition is found where a particular allele results in the production of a non-functional protein, i.e. a null allele. Homozygotes for a null allele will not produce a band on the electrophoretic gel if the staining procedure relies on the enzyme activity. Heterozygotes will show a single band of reduced intensity (figure 4.2*e*). Null alleles normally are probably only found in polyploid organisms or at loci which have been duplicated, i.e. in those organisms which have a 'spare' locus. Heterozygote patterns may not be shown in polymorphisms controlled by sex-linked loci (pp. 50 and 118).

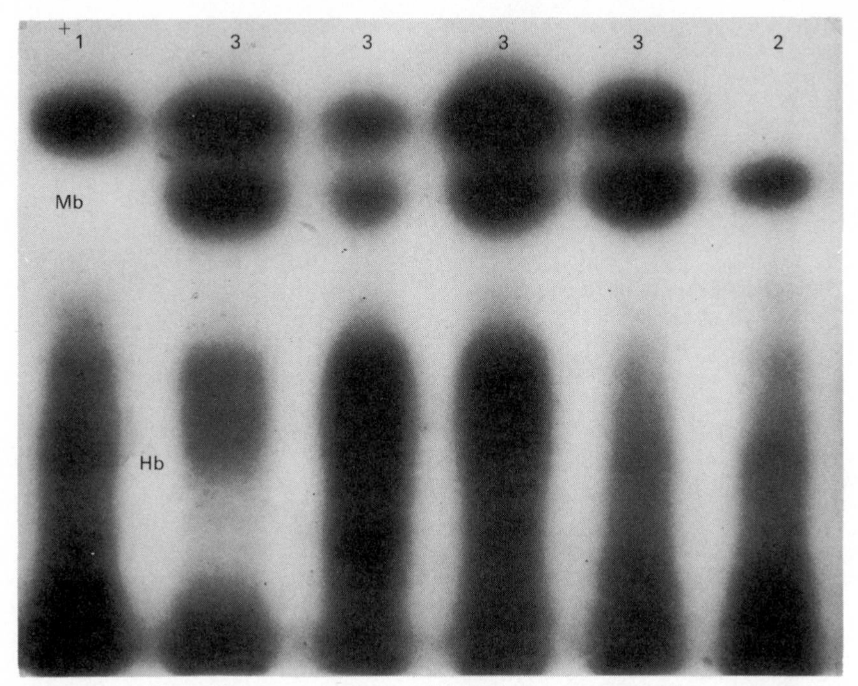

Figure 4.3. Electrophoretic patterns given by horizontal starch gel electrophoresis of heart extracts of the freshwater fish *Coregonus autumnalis pollan*. Following electrophoresis, the gel was stained for haemoglobin (Hb) and myoglobin (Mb). There are three myoglobin phenotypes as a result of a diallelic polymorphism at this locus. Three systems of nomenclature are commonly used for such alleles and phenotypes. The homozygote pattern 1 can be designated the AA type, the fast (F) type, or the 100/100 type. Similarly homozygote pattern 2 can be referred to as the BB type, the slow (S) type, or the 90/90 type, indicating that its relative electrophoretic mobility is $\frac{90}{100}$ in respect of the other homozygote. The heterozygote pattern 3 is thus the AB type, the fast/slow (F/S) type, or the 90/100 type. Naming of bands is normally carried out from the anodal side of the gel and in decreasing order of electrophoretic mobility.

Different electrophoretic forms of an enzyme which are the products of alternative alleles segregating at a locus within a species are called *allozymes*. Allozymes should not be confused with isozymes which are alternative forms of an enzyme produced by different loci. Thus each isozyme can potentially exist in a number of allozyme states. A five-banded lactate dehydrogenase zymogram can be produced in two ways: either by two polypeptides being produced in a tissue by two active loci, or by a single heterozygous locus. In a tissue where two loci are active and one is polymorphic, fifteen different lactate dehydrogenase tetramers are possible; where both are polymorphic 35 tetramers are potentially produced. The situation is even more complicated in the tetraploid salmonid and cyprinid fish, where each locus is potentially duplicated (p.

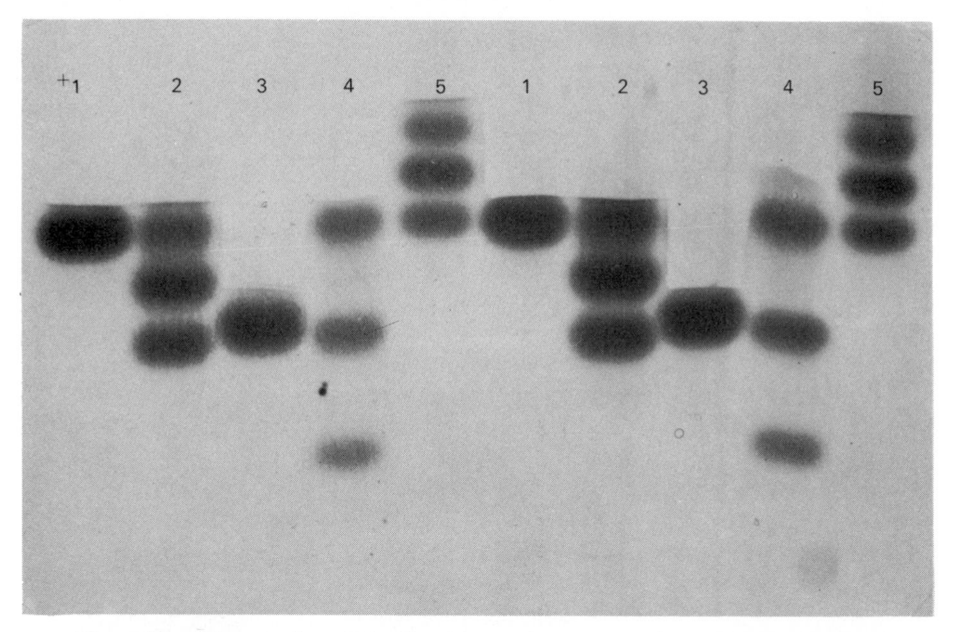

Figure 4.4. Polymorphism of the dimeric enzyme phosphoglucose isomerase in muscle extracts of the Atlantic eel *Anguilla anguilla*. Electrophoresis was carried out using a vertical starch gel system at pH 8.6. Pattern 1 is the BB homozygote; 2 the BC heterozygote; 3 the CC homozygote; 4 the BD heterozygote; and 5 the AB heterozygote. In heterozygote patterns three bands are present due to the formation of two homodimers and a heterodimer. A second locus is weakly expressed in this tissue, but for simplicity the PGI fractions produced by this locus have been eliminated by removing the agar overlay.

127). Some workers group allozymes and isozymes as defined here, under the same heading of isozymes, in some cases referring to allozymes as segregating isozymes.

Classical and balance theories of population genetics

Until about twenty years ago it was thought that most individuals within a species had the same wild-type allele present in the homozygous state at virtually all loci, and that only the occasional locus was heterozygous for the wild and a mutant allele. Under this classical model mutations were thought to be constantly introduced into the population, but most of these were assumed to be deleterious and subject to removal by natural selection. Rarely, a 'superior' allele would be produced which would increase in frequency until it became the new wild-type (figure 4.6*a*).

On the other hand, according to the balance theory there is no wild-type allele, but a number of alleles is present at most loci and any individual is regarded as being heterozygous at a large proportion of its

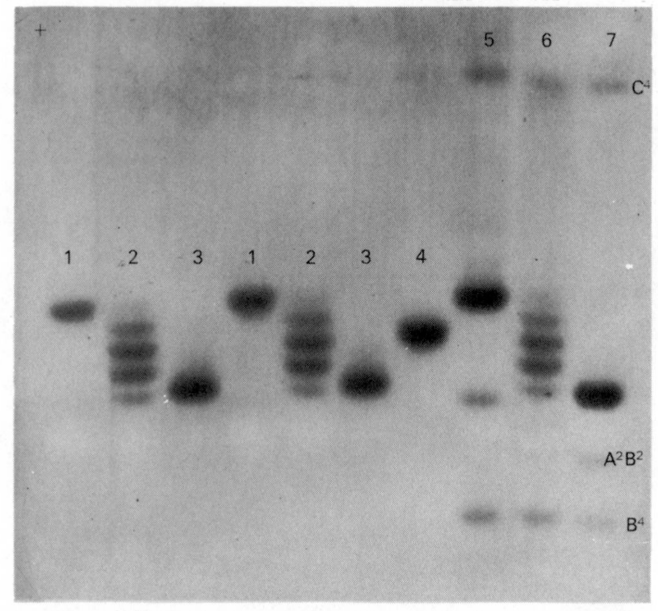

Figure 4.5. Lactate dehydrogenase allozymes and isozymes in sticklebacks. Polymorphism of the tetrameric enzyme lactate dehydrogenase is shown in muscle extracts of the nine-spined stickleback *Pungitius pungitius*.

Pattern 1 is a MM homozygote; 2 a NN homozygote; and 3 a MN heterozygote. (A and B are not used for designation of alleles to avoid confusion with LDH locus terminology,) A five-banded heterozygote pattern is given due to the formation of two homotetramers and three heterotetramers (p. 52).

Pattern 4 is the monomorphic muscle LDH of the three-spined stickleback *Gasterosteus aculeatus*.

Patterns 5, 6 and 7 are of head extracts of nine-spined sticklebacks. In these patterns, as well as the polymorphic muscle locus (A), LDH isozymes produced by two additional loci (C^4, B^4 and A^2B^2) are shown.

$$\frac{A_1\,B_1\,C_1\,D_1\,E_1\,F_1\,G_1\,H_1\,I_1\,J_1\,K_1}{A_1\,B_1\,C_1\,D_1\,E_1\,F_1\,G_1\,H_1\,I_1\,J_1\,K_2}$$

Figure 4.6a. The classical model of the genetic structure of populations.

$$\frac{A_1\,B_2\,C_1\,D_4\,E_3\,F_2\,G_1\,H_1\,I_3\,J_2\,K_5}{A_2\,B_1\,C_3\,D_1\,E_2\,F_1\,G_1\,H_2\,I_1\,J_3\,K_1}$$

Figure 4.6b. The balance model of the genetic structure of populations.

loci (figure 4.6*b*). The balance model derives its name from the fact that multiple alleles were thought to be maintained by balancing selection (p. 62).

Electrophoretic screening of many species over the past 12 years has shown that many loci are polymorphic and that a typical individual is heterozygous at a substantial proportion of its loci, in agreement with the balance model. The maintenance of the high degree of polymorphism by various forms of balancing selection, however, has not been proven.

Genetic diversity

A variety of statistics can be used to denote the amount of genetic variation in populations and species. The most informative measures are the calculated or expected frequencies of heterozygotes (heterozygosity, p. 163) and are normally expressed as the mean frequency of heterozygotes per locus ($\bar{H}_L$). Heterozygosity can also be expressed as the mean frequency of heterozygous loci per individual ($\bar{H}_I$). The values of $\bar{H}_L$ and $\bar{H}_I$ are the same, but their standard errors are different. Another commonly used measure is the proportion of polymorphic loci in a population or species (P). However, this measure does not take into account the frequency of alleles at a locus. Thus loci with alleles at 0.95 and 0.05, and 0.55 and 0.45 respectively, would both be regarded as polymorphic, but the heterozygosity (expected) in the first case is 0.095 and in the second 0.495. Other measures are the average number of alleles per locus, and the effective number of alleles per locus (p. 164).

Values of $\bar{H}$ and P are given for a range of organisms in table 4.2. The values given should not be taken as definitive statements of the amount of variability, merely as indications. Inconsistencies arise from the efficacy of different experimenters, e.g. the type of electrophoretic technique used, and whether one or more buffer systems were tried. There is also variation in the number of populations of a species which were tested. Where possible the figures used are the mean values per population, rather than the overall species value, as these two estimates can be markedly different, e.g. in the frog *Acris crepitans* only 12% of the loci were found to be polymorphic per population, but 60% of loci showed significant variability over the entire species (Dessauer and Nevo, 1969).

It can be seen from table 4.2 that, in many organisms, upwards of 30% of the structural loci are polymorphic, and individuals are heterozygous at about 10% of their loci. Since electrophoresis detects only about 30% of amino-acid substitutions (p. 42) many, if not most, loci are probably polymorphic.

It has become customary to refer to electrophoretically distinct forms of a protein as the products of alleles. However, using a range of buffer systems and electrophoretic techniques which examine different parameters of the protein molecule, and thermal and chemical stability of allozymes, it has been shown that many so-called alleles are in fact

Table 4.2 Levels of genetic variation in populations of plants and animals.

Taxa	*Ref-erences*	*Number of species*	*Mean Number of loci examined per species*	*Polymorphic per population**	*Mean hetero-zygosity*
Plants	1	8	8	0.46	0.17
Phlox drummondii	10	1	26	0.26	0.04
Invertebrates					
Drosophila	1	28	24	0.53	0.15
marine species	1	9	26	0.58	0.15
land snails	1	5	18	0.44	0.15
Theba pisana	16	1	18	0.44	0.10
Crassostrea virginica	13	1	31	0.42	0.12
Limulus polyphemus	6	1	25	0.25	0.06
Tridacna maxima	19	1	30	0.63	0.20
Otiorhynchus scaber	9	1	24	0.83	0.31
Fish	1	14	21	0.31	0.08
Rhinichthys cataractae	14	1	21	0.15	0.05
Pleuronectes platessa	5	1	46	0.48	0.12
Zoarces viviparus	4	1	32	0.28	0.09
Lepomis macrochirus	19	1	15	0.12	0.03
Amphibians	1	11	22	0.34	0.08
Bufo americanus	12	1	14	0.26	0.12
Bufo viridis	15	1	26	0.47	0.14
Hyla arborea	15	1	27	0.43	0.07
Rana ridibunda	15	1	27	0.38	0.07
Pelobates syriacus	15	1	32	0.09	0.02
Reptiles	1	9	21	0.23	0.05
Birds	1	4	19	0.14	0.04
Coturnix coturnix	3	1	28	0.57	0.17
Coturnix pectoralis	3	1	36	0.19	0.04
Phasianus colchicus	11	1	31	0.32	0.09
Mammals					
rodents	1	26	26	0.20	0.05
Rattus norvegicus (wild)	18	1	25	0.32	0.09
R. norvegicus (inbred)	18	1	25	0.12	0.01
Thomomys bottae	17	1	23	0.33	0.09
Homo sapiens	7	1	71	0.28	0.07
Mean of above			26	0.35	0.10

* In some cases the criterion for polymorphism is $\leqslant 0.95$, in others $\leqslant 0.99$.

References:

1. Selander, 1976.
2. Ayala *et al.*, 1974.
3. Baker and Manwell, 1975.
4. Frydenberg and Simonsen, 1973
5. Ward and Beardmore, 1977.
6. Selander *et al.*, 1970.
7. Harris and Hopkinson, 1972.
8. Selander *et al.*, 1969.
9. Suomalainen and Saura, 1973.
10. Levin, 1977.
11. Lucotte, 1977.
12. Guttman, 1975.
13. Longwell, 1976.
14. Merritt *et al.*, 1978.
15. Nevo, 1976.
16. Nevo and Barr, 1976.
17. Patton and Yang, 1977.
18. Eriksson *et al.*, 1976.
19. White, 1978.

groups of isoalleles which produce proteins of the same electrophoretic mobility under a single set of electrophoretic conditions (reviewed by Johnson, 1977). In a study of xanthine dehydrogenase in *Drosophila pseudoobscura*, by the serial use of four different electrophoretic approaches coupled with heat stability, Singh *et al.* (1976) found 37 alleles where previously only 6 'alleles' had been suspected. King and Ohta (1975) have pointed out that electrophoretic bands should only be regarded as electromorphs since the same phenotype may result from two or more alleles.

In electrophoretic surveys of most species, 20–100 loci have generally been examined. This is a small proportion, perhaps 1 in 1000 of the structural loci in the average organism and may represent only 0.0005% of the total DNA (Powell, 1975). Most of the loci examined are those that code for enzymes and other soluble proteins. Loci coding for non-soluble proteins and regulatory loci are not included in the survey. It is therefore necessary to question whether the loci examined by electrophoresis are typical of the genome as a whole. The loci examined are to an extent random in that they are chosen solely on the basis of a staining method being available for the protein which they produce.

Notwithstanding the problems already mentioned, a number of general features on protein polymorphism are worth noting. There is considerable variation among even closely related organisms in the extent of polymorphism. Vertebrates appear to have a lower level of polymorphism than invertebrates. Even supposedly pure lines of laboratory rats (Eriksson *et al.*, 1976) and self-fertilizing plants (Allard and Kahler, 1972) show a considerable degree of heterozygosity. An exception is a self-fertilizing snail *Rumina decollata* which was introduced into the United States and in which no genetic variation has been detected (Selander and Kaufman, 1975) although French populations show a 'normal' degree of variability. Genetic variation has likewise not been observed in the elephant seal *Mirounga angustirostris* (Bonnell and Selander, 1974).

There is also extensive variation among proteins in the frequency of polymorphism. Enzymes such as esterases have been found to be polymorphic in a high proportion of organisms which have been examined, whereas cytochrome c is invariably monomorphic. The existence of a large number of polymorphic loci means that there are many potential genotypes in each species. If all alleles are in linkage equilibrium, then the probability that two genomes, one from each of two randomly chosen conspecific individuals, have the same set of alleles for 30 000 loci is $(1-\bar{H})^{30\,000}$ which is equal to 10^{-1372} when $\bar{H} = 0.1$ (Nei, 1975). Thus within the vast majority of species no two individuals are genetically identical, except monozygotic twins.

Theoretical population genetics

If two alleles (A and B) are segregating in a population, we can represent the frequencies of these alleles respectively by p and q such that $p+q = 1$. If the population is an outbreeding sexually-reproducing one, two types of male gametes (A and B) and two types of female gametes (A and B) will be produced. These will form three types of zygotes (AA, AB, BB) in the ratio 1:2:1. Since the frequency of allele A is p, the frequency of genotype AA will be p^2; likewise $2pq$ for AB and q^2 for BB. Thus if allele A has a frequency of 0.4, the expected genotypic frequency of the AA homozygote will be 0.16. This mathematical relationship was discovered independently in 1908 by the English mathematician G. H. Hardy and a German physician W. Weinberg, and has become known as the Hardy-Weinberg Law. A population which obeys this law is said to be in Hardy-Weinberg equilibrium.

In many cases the genetic basis of protein polymorphisms has not been established by breeding experiments. The agreement of observed results with that expected from the Hardy-Weinberg Law has frequently been taken as an indication of the validity of the genetic hypothesis.

The Hardy-Weinberg Law applies to an equilibrium population only. In general, five forces can be considered as causing populations to deviate from equilibrium.

mating choice
mutation
migration
genetic drift
natural selection.

Mating choice. The Hardy-Weinberg Law applies only to an outbreeding sexual population which is panmictic, i.e. every individual has an equal chance of mating with every other individual of the opposite sex.

Mutation. Mutation introduces new variation into populations and brings about changes in allelic frequencies. However, for all practical considerations of deviations from Hardy-Weinberg equilibrium, the rate of mutation is sufficiently low to be neglected as a modifier of allelic frequencies.

Migration. The interchange of individuals between populations can change the frequencies if disproportionate numbers of individuals carrying a particular allele enter or leave the population. Differential migration is called *gene flow*.

Genetic drift. Genetic drift is a random process which operates especially in small populations. All the stages in the life cycle of organisms, from fusion of gametes through development to reproduction, are profoundly influenced by chance. In any generation, the frequency of an allele will be distributed about a mean value p with a standard error of

$$\sqrt{\frac{p(1-p)}{2N}}$$

where N is the number of breeding individuals in the population. As the population size increases, the standard error of p will decrease. This fluctuation from generation to generation is called *genetic drift*. In small populations allelic frequencies will drift with time, and alleles may be lost, with consequent decrease in genetic variability. An important aspect of this is that the allelic frequencies of a number of small populations formed from a single large one will diverge due to genetic drift. A population which passes through a period of reduced numbers (bottleneck) will lose some of its alleles and will take a long time to recover to the original level—of the order of the reciprocal of the mutation rate, i.e. about one million years (Nei, 1976). It has been suggested that the variation among species in the extent of polymorphism is a reflection of the time which has passed since the last bottleneck. A special type of drift is the *founder effect*, where founder individuals of a new population are few in number and represent only a limited part of the variation present in the parental population.

Natural selection. If an allele renders its carriers more likely to reproduce, it will place those individuals at an advantage relative to other members of the population. Over generations this allele will increase in frequency, i.e. in evolutionary terms, it will be selected for. Conversely, a disadvantageous allele will be selected against and eliminated from the population.

Of these various forces which change the genetic make-up of a population, natural selection is the only inherently directional process. Migration and genetic drift are random with respect to the genetic composition of a population, and are functions of the population structure. Selection can also operate independently on each allele, whereas migration and genetic drift affect the entire genome uniformly.

Biological significance of polymorphism

The discovery of the high level of genetic variability in natural populations has raised the question of the adaptive significance of this variation. A new mutant allele can have one of three effects on its carrier:

1. It may function in superior fashion and lead to increased reproductive fitness of the carriers, in which case it will be selected for.
2. It may render the carriers less reproductively fit and be selected against.
3. It may not affect the reproductive fitness of its carriers and its fate will depend on the balance between genetic drift and its reintroduction by mutation. In selective terms it is neutral.

In small populations, even if selection is fairly strong, allelic frequencies can change as a result of genetic drift. An allele's fate will be decided primarily by genetic drift when the coefficient of selection is much less than $1/N_e$ where N_e is the effective population size (number of breeding adults). Thus a mutant allele which is subject to selection in a large population may become neutral in a small population.

Selectionist versus neutralist hypotheses

While all biologists are agreed that some alleles are inferior, there is controversy as to whether the almost universal occurrence of polymorphic loci is the result of selection or not. The hypothesis that protein polymorphisms and molecular evolution are due to genetic drift of selectively neutral mutants was proposed by Kimura in 1968, and King and Jukes in 1969. This view, which has been termed non-Darwinian evolution,* is in sharp contrast to the selectionist or balance (Neo-Darwinian) concept of genetic variation. Many studies attempting to resolve this conflict have been published over the past ten years. These have involved experiments to examine functional variation among electromorphs; possible correlations between the patterns of allelic frequencies and environmental variables such as temperature, exposure, and latitude; and correlations between allelic frequencies, niche variation and spatial distribution of populations. The observed allelic frequencies have been compared also with those predicted from mathematical models. It is not appropriate in this book to review the several thousand papers which have been published on this topic, and the reader is referred to the reviews listed in Further Reading for comprehensive treatments. However, some selected evidence and theories will be presented, in so far as they relate to later chapters.

A new allele which is advantageous to the species can completely replace the original allele or can co-exist with it. This leads correspondingly to two types of polymorphism: transient and balanced. Transient polymorphism exists when a new superior allele is replacing the original one. Since this is a very slow process requiring several hundred generations, both alleles will be present, i.e. a polymorphism will exist throughout this time. Transient polymorphisms also occur when neutral mutations increase in frequency by genetic drift, i.e. polymorphism is a phase of allelic substitution. A balanced polymorphism occurs when two or more alleles are maintained in a species by selection. This can be due to selection for superior heterozygotes (*heterosis* or *overdominance*) or to selection for different alleles in different parts of the range, different times

* As Grant (1977) has pointed out, the term non-Darwinian evolution is a misnomer as Darwin himself recognized the possibility of selectively neutral polymorphism.

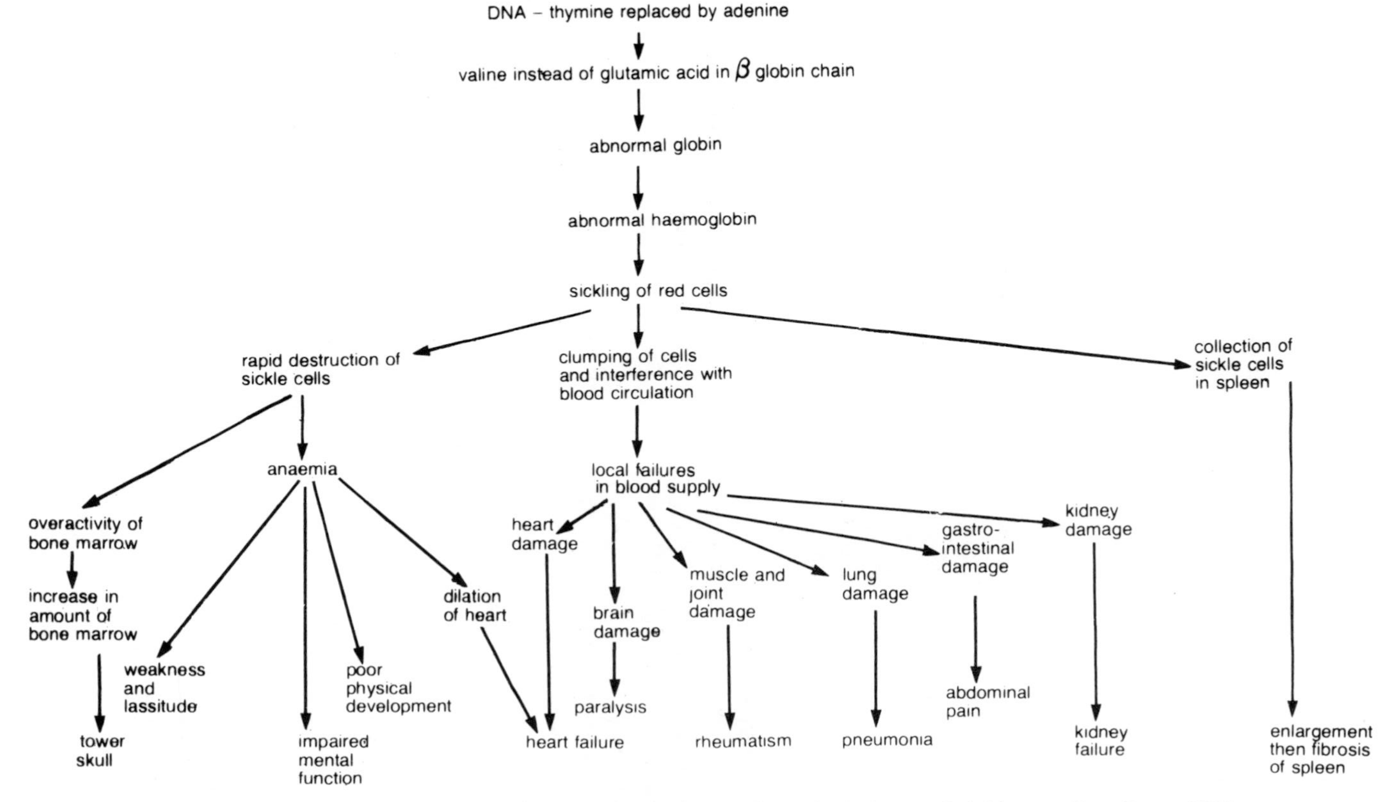

Figure 4.7. Phenotypic effects as the result of a single base change in the human β globin gene (from Berry, 1977).

of the year, or different population densities (*frequency-dependent selection*).

The first discovered example of an overdominant protein polymorphism, and still probably the most conclusive one, is that involving sickle-cell haemoglobin. A polymorphism exists at the β haemoglobin locus involving allele A, which produces the normal β polypeptide, and allele S. The polypeptide produced by the S allele differs from the normal polypeptide in that at position 6 the amino acid valine is substituted for glutamic acid. This means that the $\alpha_2\beta_2^S$ molecule has a decreased electric charge compared with that of the normal $\alpha_2\beta_2^A$ haemoglobin and so the two forms are electrophoretically distinguishable. The solubility of the sickle-cell haemoglobin is normal in oxygenated form, but drastically reduced in deoxygenated form, causing it to precipitate in the erythrocytes. When this happens, the erythrocytes become deformed and assume a characteristic sickle shape. This single amino-acid change as the result of a base substitution has a profound effect on the overall phenotype of the individual (figure 4.7).

The homozygous condition for the sickle-cell allele is usually lethal, and individuals do not survive to reproduce. In spite of this strong selection against homozygotes, the sickle-cell allele is present at frequencies of up to 0.2 in certain parts of Africa. It has been shown that sickle-cell heterozygotes enjoy some protection against *falciparum* malaria. The moderate frequency of the sickle-cell allele thus results from a balance between selection against the homozygote and selection for the heterozygote.

Power (1975) has presented a model, based on reduction of parasite populations by sickling in certain parts of the body, of how sickle-cell haemoglobin produces resistance to malaria. Clearly this is an example of a balanced polymorphism, but is it representative of the way in which most polymorphisms are maintained? The fact that it is still, after 20 years, one of the few clear-cut examples is cited by neutralists as evidence for the rarity of selectively maintained polymorphisms.

In determining selection at the molecular level, it is virtually impossible to separate selection at a locus from selection on groups of linked genes which have one advantageous locus–the 'hitch-hiking' effect (Ohta and Kimura, 1971).

Functional differences in electromorphs

If the different protein morphs produced by alternative alleles at a locus are selectively maintained, then they should differ in their ability to carry out their normal function. A number of attempts have been made to determine if functional differences exist among allozymes and other protein variants. However, in most cases the differences upon which selection acts are probably relatively small, and less than the experimental error involved in the techniques for *in vitro* measurement of enzyme activity and kinetics.

An important study is that by Koehn (1970) on the sucker fish *Catostomus clarki*, which lives in the Colorado River. This river flows from the Rocky Mountains through Arizona to the Gulf of California. The temperature of the upper reaches is around 0°C, whereas in the lower river it is 20-25°C or even higher in the summer. In the most anodally migrating esterase fraction of the serum of this fish, Koehn observed a polymorphism which appeared to be due to the segregation of two co-dominant alleles, i.e. a pattern similar to that of figure 4.2*a*. Allele A was at its greatest frequency in the southern populations, but B was absent in the extreme south. The more northerly populations had a decreased frequency of A until, at the northern end of the range, it was reduced to 0.18. This clinal pattern over the geographical range is, in itself, indicative of the variation being maintained by selection in response to an environmental parameter. The most obvious factor in this case is temperature.

In one of the tributaries where the population was very small and had been isolated from other populations for at least 10 000 years, the frequency of the alleles was that to be expected from its geographical position. Due to genetic drift in a small population, this would be unlikely to be the case if selection were not maintaining it at these frequencies. Koehn isolated the three esterase allozymes and determined their hydrolytic activity in relation to temperature. At 37°C the AA homozygote exhibited its greatest activity, while the activity of the BB types was nearly zero. As the temperature decreased, so did the activity of the AA homozygote, but the activity of BB type increased until at 0°C it showed its maximum activity, which was nearly 10 times that at 37°C (figure 4.8). The heterozygote AB exhibited a maximum activity over a wide range of intermediate temperatures, but the activity decreased markedly at the extremes. Here then there is a functional explanation for the observed pattern of geographical variation. The esterase allele (A) which was observed at highest frequency in populations at the southern part of the species range produces in the homozygous state an enzyme which has its maximum activity at the highest temperature. Conversely, allele B produces an enzyme adapted to cold conditions. The results which Koehn obtained seem at first sight to provide an excellent example of a balanced polymorphism. However, Powers and Powers (1975) have suggested that because of several factors, including the differences in molecular weights of esterase allozymes, the non-specificity of esterase reactions, the secondary esterase activity of several enzymes with other primary activity and the way in which the assay was carried out, the results are less conclusive than they appear.

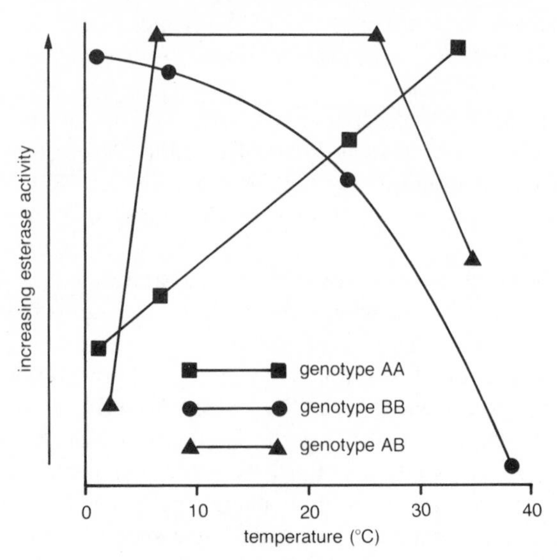

Figure 4.8. Relationship between esterase activity and temperature for three esterase genotypes of *Catostomus clarki* (redrawn from Koehn, 1970).

In the blood and egg white of birds there in an iron-binding protein, transferrin, which has been found to be polymorphic in virtually all species of birds examined in sufficient numbers. In the pigeon *Columba livia*, the two homozygotes differ by a single amino-acid substitution of serine for aspartic acid/asparagine (Frelinger, 1973). Egg or ovo-transferrin inhibits growth of iron-dependent organisms such as certain yeasts and bacteria. Heterozygous transferrin of *Columba livia* (Frelinger, 1972) and of the ring-necked pheasant *Phasianus colchicus* (Lucotte and Kaminski, 1976) has a greater inhibitory effect on the growth of yeast *Saccharomyces cerevisiae* than either of the homozygous transferrins. Since the ovo-transferrin genotype of the egg is determined by the genotype of the female, heterozygous females hatch a greater percentage of their eggs than do homozygous ones. This greater fecundity of heterozygous females may account for the maintenance of transferrin polymorphism in birds.

Activity and kinetic differences among allozymes at different temperatures have been shown for lactate dehydrogenase in the minnow *Pimephales promelas* (Merritt, 1972) and in the killifish *Fundulus heteroclitus* (Powers and Powers, 1975) and for alcohol dehydrogenase in *Drosophila melanogaster* (Day *et al.*, 1974).

Polymorphism may serve as an important mechanism for broadening the environmental tolerance of a species. Since most enzymes can operate optimally only within a narrow range of environmental factors, such as temperature, and since heterozygotes may show a wider tolerance range, it is to be expected that ectothermic animals such as inverterates and fish should show a higher genetic heterozygosity than do higher vertebrates which have physiological and behavioural homeostatic mechanisms. This suggestion is borne out by the information presented in table 4.1. A corollary of this is that organisms living in variable environments should have a higher heterozygosity than related species which live in more stable environments. Comparison among various species of inter-tidal and infaunal molluscs (Levinton, 1973; Wilkins, 1975) and among four species of Israeli toads and frogs (Nevo, 1976) which live in varying unpredictable conditions, have suggested that heterozygosity is positively correlated with environmental heterogeneity.

Inter-population comparisons

Samples of a species taken from different areas may differ significantly in their allelic frequencies. This could be due to selection for different homozygotes under varying environmental conditions (diversifying selection) or to genetic drift in isolated populations. In a study of five alleles at a leucine aminopeptidase locus in the mussel *Mytilus edulis*, Murdock *et al.* (1975) found that two populations only 100 m apart had quite different allelic frequencies, while some widely separated populations (350 km) had almost identical allelic frequencies (figure 4.9). In this case a significant correlation was found between allelic frequency and the relative amount of wave action (exposure) at the sites investigated. Since individual female *Mytilus* produce some 2.5×10^7 gametes per season, there is ample opportunity for differential survival.

In a widespread panmictic population, selection may produce differential survival in different regions and result in allelic frequency variation. All individuals of the American eel *Anguilla rostrata* spawn in one region. However, Williams *et al.* (1973) found that in samples from the eastern coast of North America the most common allele at each of three loci varied among localities in a significant linear fashion in relation to latitude. Since these differences among localities must arise by divergence from a common gene pool of zygotes, differential survival in response to an environmental variable which changes with latitude (e.g. temperature, duration of planktonic life) seems a likely explanation.

In the wild oat *Avena barbata*, Clegg and Allard (1972) have found that the genotypic frequencies for certain enzyme loci are strongly correlated with the humidity of the environment. However, Lewontin (1974) contends that, in the absence of recombination due to self-fertilization, this is probably the result of a linkage effect with one or more unknown loci under selection.

Under conditions of diversifying selection, heterozygotes may be produced by migration between populations. Such heterozygotes may

Figure 4.9. Geographical variation in allelic frequencies at the leucine aminopeptidase-2 locus in Irish populations of *Mytilus edulis*. Allelic frequencies in each population are proportional to the area of the circle occupied by its symbol (from Murdock *et al.*, 1975).

have a lower fitness than the homozygotes, and this 'negative heterosis' may promote speciation (Manwell and Baker, 1970).

When a polymorphism is maintained by overdominance or some other forms of balancing selection, similar allelic frequencies should be found in geographically isolated populations of a species. A number of studies have found this to be the case, and it is suggested that such similarities would be unlikely to result from random events. Kimura and Ohta (1971) have argued that occurrence of similar allelic frequencies in different populations is not contrary to the neutrality, since a relatively small amount of migration between populations may serve to equalize frequencies. (A population is effectively panmictic if Nm is greater than 1, where N is the number of breeding individuals and m is the migration rate per generation). Limited migration is probably a feature of many animal populations. Higher vertebrates can move at will over large distances, and invertebrates can be passively transported. However, in the case of freshwater fishes it is often possible to state that gene flow is absent between populations in geographically isolated lakes. In the pollan *Coregonus autumnalis*, the same frequencies for the two alleles controlling a myoglobin polymorphism have been found in two Irish lakes and in a sample from Alaska, all three populations having been isolated for at least 10 000 years (Ferguson, 1975, and unpublished). It is possible, of course, that it is only the electromorphs which are the same and that the actual alleles are different (p. 59).

Aspinwall (1974) has studied an interesting case where gene flow is also known to be absent. He examined 32 populations of the North American pink salmon *Oncorhynchus gorbuscha*. In common with the Atlantic salmon, the pink salmon is anadromous but, unlike the former, it has a rigid life cycle, returning to the spawning streams at the exact age of two years (± 10 days). Due to this fixed schedule, two genetically isolated populations, called odd-year and even-year, exist in the same stream in many parts of Alaska and British Columbia. Allelic frequencies at three loci showed considerable uniformity, either within distantly spaced odd-year or within even-year Alaskan populations. However, differences in allelic frequencies occurred between odd-year and even-year populations occupying the same streams. Since populations in the same stream are assumed to be under the same type of selection, Aspinwall argued that these results support the neutral mutation/random drift hypothesis. The uniformity within either type is consistent with this hypothesis, as substantial gene flow has been shown between separate river systems.

Genetic co-variation between species

The same polymorphism may be present in two closely related species as a result of its presence in a common ancestor. If there is geographical variation in the allele frequencies at such a locus, then it should vary concomitantly in both species if governed by selection. Harrison (1977) has shown that two sibling species of crickets (*Gryllus veletis* and *G. pennsylvanicus*, which do not interbreed) show a parallel pattern of geographical variation in their allelic frequencies at the phosphoglucose isomerase locus. Parallel changes have been found also in the marine molluscs *Modiolus demissus* and *Mytilus edulis* (Koehn and Mitton, 1972).

Deviations from Hardy-Weinberg expectations

If selection is taking place, then the observed genotypic frequencies in a population should deviate significantly from those predicted by the Hardy-Weinberg law. However, large sample sizes and/or strong selection are necessary for such deviations to show up with normal sampling procedures. A number of factors other than selection can produce deviations from Hardy-Weinberg expectations, e.g. treatment of two fully or partially isolated populations characterized by different allelic frequencies, as a single panmictic population can show a significant deficit of heterozygotes. Sex-linkage can produce an apparent deficit of heterozygotes. A difference in allelic frequencies between the sexes and disassortative mating (pairing of unlike individuals) can generate an excess of heterozygotes. Several studies of natural populations have demonstrated both heterozygote deficits and excesses and have been used, where the above factors can be discounted, to suggest that selection is taking place.

The eelpout *Zoarces viviparus* has been extensively studied by Christiansen, Frydenberg and others in Denmark. Since this fish is live-bearing, the possibility of obtaining mother-offspring combinations makes it particularly suitable for detailed investigations of selection. A significant deficit of heterozygotes for an esterase polypmorphism has been found among adult eelpouts and is not found in embryos. This deficit in adults has been shown (Christiansen, 1977) to arise from selection occurring by differential survival of young fish in the period from birth to maturity at the age of two years.

Genetic load

One of the arguments put forward against the selective hypothesis is that, if polymorphisms are maintained at many loci by balancing selection, then this would cause too large a genetic load in the population. If one genotype at a locus is selectively superior, others must be inferior. When taken across thousands of loci (since alleles at many loci segregate independently), few, if any, individuals will have the superior genotype combination at all of them. In this event it has been argued that the cumulative reduced fitness or genetic load would be greater than the reproductive capacity, and so most polymorphisms must be selectively neutral.

In the main, however, selection does not operate on individual loci, but rather on the overall genotype which manifests itself in the phenotype of the individual. Allelic changes at a single locus may have relatively little effect on the final phenotype, except for a small proportion of loci which may produce large phenotypic changes and be subject to strong selection.

If single-locus allelic changes are too small to measure, and if a large number of polymorphisms are maintained by overdominant selection, then phenotypic differences should be detectable between individuals with different degrees of total heterozygosity. In the oldfield mouse *Peromyscus polionotus*, higher heterozygosity has been found to confer superior fitness (Smith *et al.*, 1975). Using the same species, Garten (1977) has carried out behavioural experiments and has found that the mean latency to enter the 'open field' (novel brightly lit arena) decreased, and mean ambulatory behaviour increased with increasing mean genetic heterozygosity. This might imply an increased exploratory behaviour by individuals of higher heterozygosity, and this could be advantageous in conditions of food scarcity.

Constancy of protein evolution

One of the strongest pieces of evidence in favour of the neutral hypothesis is the remarkable uniformity in the rate of evolution of particular proteins throughout a wide range of organisms (chapter 8). If these changes were occurring under selection, it is to be expected that homologous proteins in different metabolic and environmental surroundings might evolve at varying rates. It has been pointed out that the rate is only constant when changes are measured over a long period of time, which damps down relatively short-term irregularities in evolutionary rate. It has been shown also that proteins, or parts of proteins which are subject to less functional constraints, incorporate amino-acid substitutions at a faster rate.

Conclusions

Much of the evidence supporting one or other theory is equivocal and subject to interpretation in favour of either viewpoint. The argument, however, seems to be not whether protein polymorphisms are maintained by selection *or* genetic drift, but rather which is the more common and important process in molecular evolution. Clearly some polymorphisms are adaptively significant and under strong selection, while others may be less important and functionally neutral or almost so. Any mutation may range in its effects from distinctly advantageous through almost neutral, neutral, slightly deleterious, to distinctly deleterious. In an expanding post-bottleneck population, even slightly deleterious mutants may be selectively neutral. In changing environmental conditions, an allele may be neutral when it arises but be subject to strong selection at a later time. Thus, selectively advantageous, neutral and disadvantageous, are concepts relative only to particular environmental conditions. Manwell and Baker (1970) have suggested that organisms with greater heterozygosity may be 'pre-adapted' to novel environmental circumstances in that they are more likely to have some alleles which can cope under the new conditions.

Just as the nature versus nurture controversy was solved by the quantitative concept of heritability, so the present random versus deterministic argument may be settled by a quantitative assessment of the relative contribution of each in molecular evolution.

A complication which arises, however, is that some of the evidence on which the debate has centred so far may be invalid due to the limitations of the electrophoretic techniques. As already mentioned, a single electromorph class may be produced by a number of different alleles*. So similar allelic frequencies in geographically isolated populations may merely reflect similar electromorph frequencies produced by unique alleles in the two populations. With the development of nucleic acid sequencing, a new set of data is emerging which may help to resolve the situation. This information has enabled the rate of mutations occurring at the third codon position (about two-thirds of which are synonymous with regard to amino-acid coding) to be determined. It seems likely that such synonomous mutations are selectively neutral, since they produce the same protein phenotype. Kimura (1977) had pointed out that histone IV, which shows the lowest known rate of amino-acid substitutions of any protein, has the highest known rate of synonymous base substitutions.

Fortunately from the standpoint of the use of protein and nucleic acid information in taxonomy, it does not matter whether polymorphisms and molecular evolution are the result of neutral changes or not. If most changes in structural loci are selectively neutral, so much the better, as the possibilities of convergence and directional changes will be minimized.

* Breeding experiments may show Mendelian segregation of electromorphs, but this does not prove that they are the products of single alleles.

CHAPTER FIVE

INTRA-SPECIFIC SYSTEMATICS

WITH THE DISCOVERY OF THE UNEXPECTEDLY LARGE AMOUNT OF genetic variation which is present within species, the need to characterize populations, races and sub-specific groups became more evident. The realization of the correlation of genetic variants with growth rate, disease resistance, chemical resistance, and other economically important traits, has also given further impetus to these studies.

Inter-population genetic variability

At each structural locus two populations of a species may be as follows:

1. Both monomorphic for the same allele.
2. Both monomorphic for different alleles.
3. One monomorphic, the other polymorphic, with or without a common allele.
4. Both polymorphic for the same alleles present at the same or different frequencies.
5. Both polymorphic for different alleles.

Several measures are available to quantify the genetic differences between populations. For monomorphic loci, a simple coefficient representing the proportion of loci with identical alleles is appropriate. Various statistics have been proposed to estimate differences based on polymorphic loci (Cavalli-Sforza, 1969; Hedrick, 1971; Rogers, 1972; Nei, 1972; Richardson and Smouse, 1976; Flake and Lennington, 1977). These make use of allelic frequencies, or genotypic (electromorph) frequencies, and/or relative electrophoretic mobility of electromorphs.

The most widely used statistics are those of Nei (p. 166). Although Nei's original arguments assumed selective neutrality, the numerical value of his coefficients are not dependent on this assumption. Nei's coefficient of *genetic identity* I ranges from zero (no alleles in common at a locus) to one (the same alleles at identical frequencies). The standard *genetic distance* D between two populations is given by

$$D = -\log_e I$$

The mean genetic identity ($\bar{I}$) and genetic distance ($\bar{D}$) are the mean values over all loci studied (including monomorphic ones).

When the number of allelic substitutions per locus follows a Poisson distribution, and when codon substitutions at a locus are independent of each other, D may be interpreted as a measure of the mean number of electrophoretically detectable substitutions per locus which have accumulated since the two populations separated from a common ancestor. The range of D-values is from zero to infinity. The sum $I+D$ is greater than one because two or more codon substitutions may have occurred at the same locus.

The actual number of codon substitutions is given by D/c, where c is the proportion of codon substitutions which are electrophoretically detectable.

If the rate of codon substitutions per year is constant, and if two populations (within an order of magnitude of the same size) are in equilibrium with respect to the effects of mutation, selection and genetic drift, then D is linearly related to the time of divergence T of the two populations such that

$$D = 2\alpha T$$

where α is the rate of electrophoretically detectable codon changes per locus per year. Nei has estimated an average value of α to be 10^{-7}, therefore

$$T = 5 \times 10^6 D$$

i.e. when $D = 1$, the populations have been isolated for approximately 5 million years (but see p. 103).

Time of divergence estimates based on D-values are in some cases in close agreement with times based on geological and other evidence. Migration of genes between the two diverging populations will reduce the genetic distance and result in an underestimate of the true times of divergence. However, even if the absolute times are not reliable, relative divergence times may be valuable in some problems.

Table 5.1 gives some examples of $\bar{I}$ and $\bar{D}$-values for populations and subspecies. Levels of genetic similarity among populations of most species are high, with an average of five electrophoretically detectable codon changes per 100 loci. Most geographically separated populations share the same alleles at monomorphic and polymorphic loci, and at similar frequencies in the latter. Only a small proportion of loci have alleles at significantly different frequencies, even fewer having unique alleles (figure 5.1). Populations sufficiently distinct to be regarded as

Table 5.1 Mean values of genetic identity ($\bar{I}$) and genetic distance ($\bar{D}$) between geographically separated populations, and between conspecific sub-species.

Taxa	*Reference*	*Geographical populations* $\bar{I}$	$\bar{D}$	*Sub-species* $\bar{I}$	$\bar{D}$
Invertebrates					
Drosophila willistoni group (5 species)	1	.968	.032	.796	.228
Limulus polyphemus	3	.990	.010		
Phoronopsis viridis	3	.996	.004		
Tridacna maxima	3	.968	.032		
Arbacia punctulata	4	.949	.052		
Vertebrates					
Centrarchidae (sunfish 19 species)	2	.977	.024	.843	.171
Taricha (salamanders–3 species)	3	.984	.017	.836	.181
Poeciliopsis occidentalis	6	.889	.119		
Thomomys bottae	5	.866	.144		
Mus musculus	7	.987	.013	.798	.226
Plants					
Avena barbata	3	.714	.336		
Lycopodium lucidulum	3	.976	.024		
Lupinus	3	.966	.034		
Hymenopappus	3	.956	.044		
Stephanomeria	3	.980	.020		
Mean of above taxa		.944	.058	.818	.201

References:

1. Ayala *et al.*, 1974.
2. Avise and Smith, 1977.
3. Ayala, 1975.
4. Marcus, 1977.
5. Patton and Yang, 1977.
6. Vrijenhoek *et al.*, 1977.
7. Britton and Thaler, 1978.

separate sub-species generally show about four times as much genetic divergence as geographically separated but morphologically similar ones. In the case of sub-species, loci are either similar in allelic content (the majority) or fixed for alternative alleles (figure 5.2).

Breeding systems

The nature and extent of genetic variation which is present within a species depends on its mode of reproduction. In outcrossing sexually reproducing organisms, a species is made up of one or more populations of genetically unique individuals, partially or completely isolated reproductively. Many organisms reproduce asexually, by partheno-genesis or by self-fertilization, giving rise to groups of individuals (clones) which, in the absence of mutation, are genetically identical to their progenitor and to each other. The individuals of a species found in a

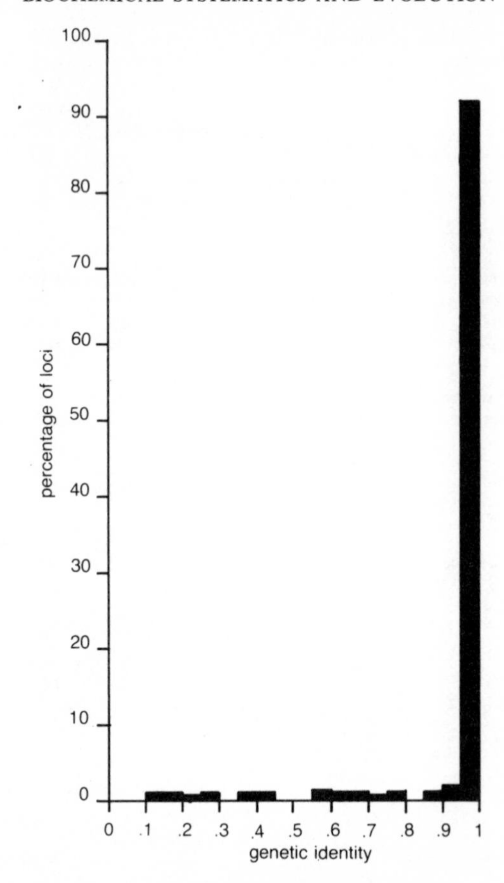

Figure 5.1. Histogram showing the percentage of loci within a given range of genetic identity among local populations of the same subspecies of sunfish *Lepomis macrochirus* (redrawn from Avise and Smith, 1977).

particular area may consist of a number of genetically distinct clones which change in relative frequency under the action of genetic drift and natural selection. When a mutation arises, the individual possessing it will form a new clone. Within a diploid clonal organism, polymorphism may exist, but individual clones are fixed homozygotes or heterozygotes at each locus.

Electrophoretic studies of genetic variation can give valuable information on the reproductive strategy practised by a species. In turn, this information is necessary for correct systematic interpretation.

In the Mediterranean area the snail *Rumina decollata* occurs as a complex of homozygous strains produced by a breeding system of facultative self-fertilization. Colonies may be composed of two or more strains (Selander and Kaufman, 1975). In one area studied in detail, the

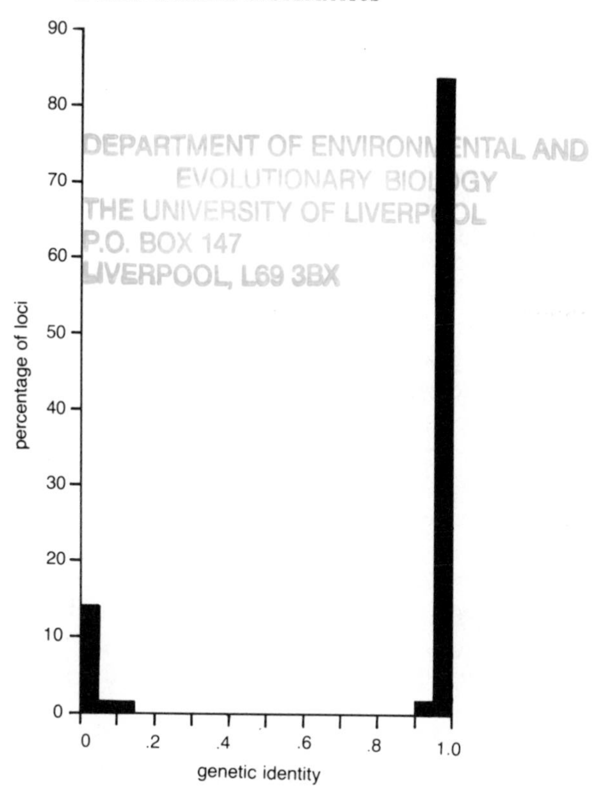

Figure 5.2. Histogram showing the percentage of loci within a given range of genetic identity among populations belonging to different subspecies of sunfish, *Lepomis macrochirus macrochirus* and *L.m. purpurescens* (redrawn from Avise and Smith, 1977).

two strains present showed the same allele at 13 enzyme loci and were fixed for different alleles at another 13. In some mixed colonies heterozygous individuals were found, showing that outcrossing does take place, and the frequency of this was estimated to be 15%. In this respect, *Rumina* is similar to self-fertilizing plants such as oats and barley, in which inbred strains of different genetic composition are found and where there is occasional outcrossing. As with *Rumina*, the various plant strains are often fixed for unique alleles and hence show unique electromorph patterns which can aid in their characterization. Electrophoretic patterns have been used to identify barley cultivars, an essential prequisite in determining varietal purity and originality in plant variety protection and other agricultural applications (Almgård and Landegren, 1974; Fedak, 1974). An index of all European potato varieties has been compiled on the basis of their unique esterase and peroxidase zymograms (Stegeman, 1977).

Studies of protein variation have confirmed the existence of asexual reproduction in a number of species, e.g. the sea anemone *Haliplanella luciae* (Shick and Lamb, 1977). Crozier (1973) studied the pattern of polymorphism at a malate dehydrogenase locus in *Aphaenogaster rudis* and found evidence that colonies of this ant are simple family units consisting of a single monogamous queen and her worker progeny.

In birds, the proteins of the egg white are determined by the maternal genotype. The discovery of incompatible genotypes of polymorphic egg white proteins has shown that certain eggs in nests of house sparrows *Passer domesticus* had been laid there by a female other than the 'owner' (Manwell and Baker, 1975). Multiple fertilizations of lobsters and fish have been demonstrated by the incompatability of the progeny with a single male genotype.

Population taxonomy

Ideally a character used for delimiting a population or any other taxonomic category should be present in all members of the population and not present in other populations, i.e. a unique fixed allele or its product. In outcrossing sexual organisms, inter-population gene flow keeps populations genetically very similar, and prevents the fixation of different alleles as in clones or strains of asexual and self-fertilizing organisms. As already noted, most differences between conspecific populations are those in allelic frequency. Such differences may be used to characterize the population as a whole, even if individuals cannot be identified. If several polymorphic loci can be combined, the possibility of characterizing a population increases exponentially with the number of loci applied. Polymorphic loci which are used in this way may best be regarded as markers in order to distinguish them from strict taxonomic characters. In some species, populations can be separated clearly by significant differences at many loci, while others show no detectable differences over a broad geographical area, despite study of numerous polymorphic loci.

As an example of the way in which protein variants are used as taxonomic characters and markers, an extreme case can be considered. If conspecific individuals in a sample from area 1 show only allele A at a particular locus, and in an equivalent sample from area 2 only allele B, then it can be argued that groups from 1 and 2 belong to two reproductively isolated populations. If gene flow exists between the two populations, then allele A would be introduced into population 2, and vice versa for allele B. Similarly it can be suggested that groups which have the same alleles, but at different frequencies, are partially isolated populations, or populations which have been isolated for an insufficient period of time for fixation of different alleles. These arguments, however, only hold if the polymorphism is selectively neutral. If not, then there is an alternative explanation for these observations. If AA homozygotes are selectively superior under the conditions experienced by group 1, and likewise for BB homozygotes in area 2, then only individuals of genotype AA may survive in area 1 and only BB in area 2. Less drastic differential survival could produce different frequencies in separate parts of the range of a panmictic population. Hence the observed differences could be due to directional selection rather than to lack of, or reduced, gene flow. It should be noted also that the reverse situation does not necessarily hold true; homogeneity over a wide geographical area does not invariably indicate a single panmictic population, as this could be due to similar balancing selection, or possibly the limitations of the electrophoretic technique as previously discussed.

Intra-specific taxonomy in fishery biology

As Everhart *et al.* (1975) have pointed out:

> Classification by species, subspecies and even races may not be fine enough for the fishery resource manager who may find that he must recognise subpopulations or strains to reach the subtle level needed for good fishery management.

The realization that the conventional species classification is too broad for fishery management, and the commercial need for a sound intra-specific taxonomy of fish species has led to considerable application of electrophoretic studies to this area. Similar arguments apply to other wildlife species which are commercially harvested, and also to species which are important from a conservation point of view. In short, detail of the population taxonomy of a species is a basic prerequistite for rational exploitation and management.

In the past, meristic characters such as number of vertebrae, fin-rays, etc., were used for the delimitation of fish stocks. Unfortunately these characters are plastic and subject to variation under changing environmental conditions, especially temperature during embryonic development (figure 5.3). By cold and heat shock treatment of embryos of Danish sea trout *Salmo trutta*, Tåning (1952) was able to produce individuals with the same number of vertebrae as found in natural stocks from Scandinavia and from the Mediterranean. Oxygen and carbon dioxide tensions during development were noted also to change the number of vertebrae in this species.

The other approach to the study of fish stocks is by the use of numbered tags which are attached to individual fish. These tags are valuable in tracing migrations and assessing survival rates. They do not, however, give direct information on population breeding structure. As

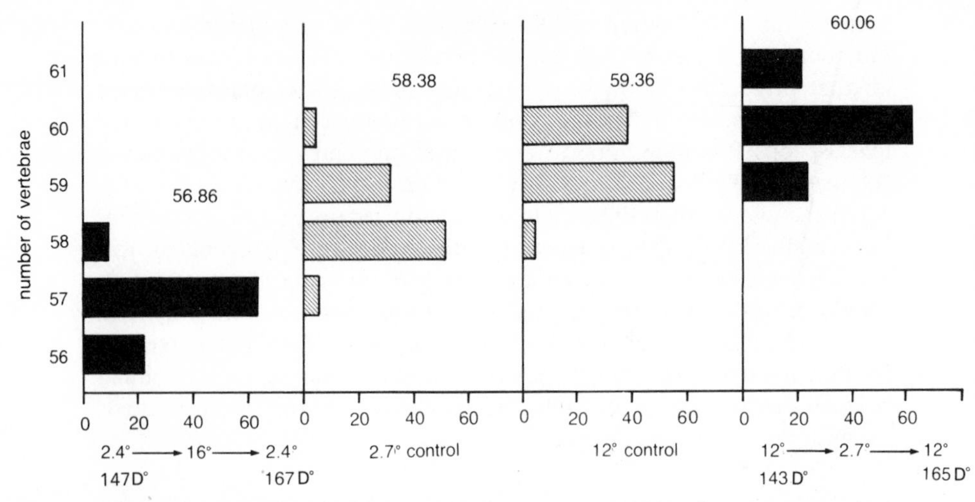

Figure 5.3. Numbers of vertebrae in offspring of sea trout subjected to temperature shock treatment during the supersensitive period of embryonic development. The extreme left figure shows the effect of heat treatment for about 20D°, and the extreme right figure the effect of cold shock ($D°$ = days × mean temperature/day). Values for untreated controls are given in the middle figures. The mean values for the numbers of vertebrae are given above the histograms (redrawn from Tåning, 1952).

noted earlier, each individual member of a sexually reproducing species is genetically unique. Thus each bears its own genetic tag, the 'number' of which is written in code—the genetic code. This code can be deciphered by examining protein variants. At the moment, techniques do not permit discrimination among individuals, but in many cases they are sufficiently sensitive to delimit populations.

Unfortunately, the use of protein markers in population studies is limited, in that many fish species have a high fecundity, and therefore considerable opportunity exists for differential survival. In some marine species, many thousands or millions of zygotes are produced, and only a small fraction of these survive to adulthood. Several well-substantiated examples of differential survival in eels and other fishes have already been noted (pp. 67–70). In some cases where allelic frequency differences in samples from geographically separate areas were originally taken as evidence of isolated populations, more detailed examination of intermediate areas has shown that a gradual change (cline) in frequency exists.

Marine fish stocks

The economically valuable sea-fishes have been subjected to detailed electrophoretic screening in an endeavour to find suitable protein markers for the delimitation of populations and other fishery 'units'. Most workers have assumed that if the observed genotype distributions in samples are in agreement with Hardy-Weinberg expectations, then differential survival has not taken place, and differences in allelic content reflect population stratification. However, large sample sizes are required before such deviations become apparent. This constraint also reduces the potential of the approach, since a reduced number of heterozygotes (compared with that expected on the basis of the Hardy-Weinberg model) may reflect either that selection is taking place or that the sample was a mixture of two or more populations characterized by different allelic frequences (Wahlund effect). If a mixed sample of roughly equal proportions taken from the hypothetical populations 1 and 2 (which are fixed for alleles A and B respectively) is treated as a single population, then alleles A and B would be found at frequencies of approximately 0.5. On Hardy-Weinberg expectations 50% ($2pq$) of the sample should be AB heterozygotes but, in fact, since two isolated populations with different alleles are involved, heterozygotes are not observed. Similar, though less dramatic, differences are found when a mixed sample is taken from two populations with the same alleles but at different frequencies. (An apparent deficit of heterozygotes can also be produced as the result of sex-linkage.)

The development in the 1960s of extensive commercial netting of the Atlantic salmon *Salmo salar* off the west coast of Greenland provided impetus to determine the relative contributions of each country to this stock, which was clearly not of local origin. Unfortunately few suitable marker loci have been found in this species. One, the transferrin (Tf) locus, has been investigated in considerable detail by Møller, Payne and others. Four alleles segregating at this locus have been found, but only one (Tf^1) is common to salmon from both European and North American waters. The Tf^2 allele is restricted to European samples, and the Tf^3 and Tf^4 alleles are exclusively North American. Payne *et al.* (1971) have suggested that the two races merit sub-specific status, namely, *Salmo salar europaeus* and *S.s. americanus.*

Differences in frequencies of the Tf^2 allele in a sample of 10 000 Atlantic salmon have been used to postulate that there are two separate races in the British Isles: a northern 'boreal' race and a south-western 'celtic' race which evolved in isolated refugia during the final phase of the last glaciation of Europe (Payne *et al.*, 1971; Child *et al.*, 1976). They suggest that when the ice covered the northern part of the British Isles, the 'boreal' race was isolated in the present North Sea area and that, when the ice retreated, the rivers in the northern parts of the British Isles were recolonized by this race. The 'celtic' race is assumed to have remained in the ice-free area to the south-west throughout this time.

Similar analysis of geographical variation in the frequency of alleles at

Figure 5.4. Variation among samples in transferrin alleles (A, B, C1, C, D) of the Atlantic cod. The sum of the shaded blocks is unity for each named site (redrawn from Jamieson, 1975).

the Tf locus in Atlantic cod *Gadus morhua* has indicated that there are several major isolated populations (Jamieson, 1975; Cross and Payne, 1978). As with the salmon, greater differences are found between eastern and western Atlantic groups than among populations in each area (figure 5.4).

The approach discussed so far equates a fish stock with a Mendelian population. Other units may be of value for management and exploitation purposes. Even if allelic frequency heterogeneity is the result of directional selection, it may be useful in the identification of separate adult stocks.

Freshwater salmonid fish

While it is to be expected that genetically isolated populations can exist within a large area such as the North Atlantic, the discovery of two or more isolated populations in small freshwater lakes is more surprising. Allendorf *et al.* (1976) have shown that two genetically isolated populations of brown trout *Salmo trutta* exist in a small Swedish lake. The two populations appear to be homozygous for different alleles at a lactate dehydrogenase locus.

In Lough Melvin (area *c.* 16 km^2) in north-western Ireland four 'morphotypes' of brown trout have traditionally been recognized by anglers. An examination of protein polymorphism in a sample of 459 trout (Mason and Ferguson, unpublished) showed a significant deficit of heterozygotes at two loci. As discussed earlier, such a deficit can occur when isolated populations having alleles present at different frequencies are treated as a single panmictic population. In this case, when the sample was partitioned into the four trout types, significant differences in allelic frequencies at these two loci were noted. Also, in each case, the samples were now in agreement with Hardy-Weinberg expectations (table 5.2). For four other polymorphic loci where no significant deficits of heterozygotes were noted, no heterogeneity in allelic frequencies was found among the morphotypes. The extent of the allelic frequency differences shown in table 5.2 suggests that there are at least three reproductively isolated stocks of trout in this lake (ferox, gillaroo, and brown + sonaghen).

In common with the Atlantic salmon, races of the brown trout probably have their origin associated with events of the last glaciation which began some 75 000 years ago and lasted until about 10 000 years ago. With the onset of the glaciation and the complete coverage of all but the extreme south of the British Isles, all freshwater fishes would have been eliminated or, in the case of the migratory salmonids, forced to retreat southwards. When the ice advanced, fish in separate geographical areas would have been driven into different refuges and isolated from one another thus allowing genetic differentiation to take place. However the glaciation was not contiuous for the 60 000 years, several retreats taking place and allowing some previously isolated populations to come into contact and interbreed. The period of isolation was probably sufficient for some genetic differentiation to occur, but not to the extent that hybrids were selected against, with the consequent acquisition of isolating mechanisms. When the ice advanced again, it seems likely that it did not follow the same pattern as previously, and so a different distribution of refuge isolation would result. These two factors—isolation for some thousands of years, interspersed with introgression between isolates—are undoubtedly responsible for the large number of intra-specific forms, not only of the brown trout but also of other salmonids such as the char (*Salvelinus* spp.) and whitefish (*Coregonus* spp.) in north-western Europe today. The several major Pleistocene glaciations which have taken place during the past million years have played a major role in the speciation and 'sub-speciation' of many groups of animals and plants.

Table 5.2 Allelic frquencies, observed (obs.) and expected (exp.) genotypic distributions, G-values (p. 164) and probabilities of deviations from expected Hardy-Weinberg distributions, for two loci in the brown trout *Salmo trutta* from Lough Melvin in western Ireland. Ferox, brown, sonaghen and gillaroo refer to types of trout recognized on morphological characters. The higher frequency allele is designated as '100'. At both loci two alleles are segregating. Data from Mason and Ferguson (unpublished).

		Phosphoglucose isomerase Locus-2							*Lactate dehydrogenase Locus-5*						
		Frequency of '100' allele		*Genotypic distributions*					*Frequency of '100' allele*		*Genotypic distributions*				
Sample	N			100/100	100/140	140/140	*G*	*P*			100/100	100/105	105/105	*G*	*P*
All trout	459	0.76	obs.	295	126	38	18.98	<.001	0.93	obs.	408	36	15	43.34	<.001
			exp.	261.5	167.5	26.4			exp.		397.0	59.8	2.2		
Ferox	42	0.96	obs.	39	3	0	0.01	>.99	0.39	obs.	5	23	14	0.93	>.3
			exp.	38.8	3.2	0				exp.	6.4	20.0	15.6		
Brown	176	0.70	obs.	92	62	22	0.60	>.3	0.96	obs.	164	10	2	5.22	>.02 <.05
			exp.	86.2	74.0	15.8				exp.	162.2	23.5	0.3		
Sonaghen	105	0.64	obs.	44	47	14	0.06	>.98	0.98	obs.	102	3	0	0.54	>.3
			exp.	43.0	48.4	13.6				exp.	100.8	4.1	0.1		
Gillaroo	136	0.93	obs.	120	14	2	2,48	>.1	1.0	obs.	136	0	0		
			exp.	117.6	17.7	0.7				exp.	136				

Isolation in refuges with different environmental and physical conditions has resulted in the evolution of a diverse range of ecological and behavioural adaptations. For example, in the salmonid fishes there are considerable differences in food preferences, growth rate, age at maturity, and time and place of spawning (autumn versus spring spawning, river versus lake spawning, inlet river versus outlet river, and so on into finer divisions). With their innate tendency to spawn in their natal areas, these different forms can exist sympatrically, but yet be reproductively isolated. This diversity of forms is difficult to deal with using conventional taxonomic categories. Following the strict biological species concept, reproductively isolated forms living in sympatry are separate species. This can lead, however, to undue splitting, and homologies between forms in different lakes are very difficult to establish. The alternative of putting them all in one species or 'super-species' is even more unsatisfactory, as it does not express the tremendous range of ecological and other adaptations found among the different forms. As previously noted, the species category is much too broad even for less variable species, since fisheries management requires an adequate taxonomy describing strains or populations which are adapted to various ecological conditions, and with appropriate productivity or angling characteristics. The requirement, then, is for a genetic catalogue of these types. Electrophoretic studies, coupled with ecological and behavioural observations, provide the most practical means of achieving this in the short time available before such genetic diversity is lost due to artificial introductions, eutrophication, exploitation and physical alteration of the habitat. The potential value of these various genotypes for future breeding programmes may not be realized until it is too late. It is the job of the systematist to highlight this genetic variation by means of an adequate intra-specific classification.

Stocking

Stocking of artificially reared individuals is a commonly used management practice with commercially important fish, molluscs and other animals. In many cases the efficacy of this procedure is unknown. If unique alleles are present in stocked individuals, they will serve as markers enabling the fate of these individuals to be followed. This procedure has been tried out with lobsters (Hedgecock *et al.*, 1976) and with rainbow trout *Salmo gairdneri*. Even if suitable markers are not present in natural stocks, then lines homozygous for different alleles can be produced by appropriate matings, or possibly by artificially induced gynogenesis or androgenesis. A danger here is that if the polymorphism concerned is maintained by balancing selection, then homozygotes may have reduced viability. Also, other linked loci may be simultaneously subjected to this artificial selection.

Since considerable heterogeneity exists among populations which make up most species, electrophoresis may be applied to determine the appropriate genotype to use for stocking a particular area. Rainbow trout from fast and slow-flowing rivers differ in their lactate dehydrogenase genotype (Northcote *et al.*, 1970). A single strain of hatchery-reared fish is unlikely to be equally successful in rivers and lakes with varying physical, chemical and biological characteristics. Stocking of four strains of rainbow trout in a Californian reservoir resulted in a harvest ranging from 2.5% to 33%, depending on the strain (Everhart *et al.*, 1975).

Breeding programmes

If fitness of a commercially important stock is taken as the magnitude of desirable traits such as fast growth rate, high fecundity, disease resistance, etc., then in general such fitness is related to the genetic variability or heterozygosity of the stock. Selective breeding programmes have shown that increasing the heterozygosity of a stock is normally beneficial, whereas inbreeding, with consequent reduction in heterozygosity, is generally harmful. For this reason, many animals and plants used in agriculture are intra- or interspecific hybrids. Electrophoretic screening of farm or natural stocks can be used to detect inbreeding and to identify genetically discrete stocks which would be likely to give increased heterozygosity on crossing. From a conservation viewpoint, it is as necessary to maintain the genetic diversity which is present within all species as it is to maintain a diversity of species. Here electrophoretic surveys can delineate populations of special interest which should be protected from artificial introductions of the same or closely related species and from environmental modifications.

Attempts to correlate protein-determined genotypes at single loci with production fitness have not been notably successful. This is probably due to most production characteristics being polygenically controlled and having a sizeable environmental component. However, alleles at the transferrin locus have been correlated with weight gain in the rainbow trout (Reinitz, 1977), milk production in dairy cows (Ashton *et al.*, 1964), embryo mortality in chickens (Gilmour and Morton, 1971), calf weight at time of weaning (Fowle *et al.*, 1967), and fertility in the skipjack tuna *Katsuwonus pelamis* (Fujino and Kang, 1968). Similar correlations have been shown for a few other loci. Correlation does not necessarily mean that these particular alleles control the desired trait, but more likely that they are linked to alleles which do. As such they are valuable markers for use in selective breeding programmes, since specific crosses can be carried out to produce desired genotypes. If, for example, AB is the optimum, then a cross of AA × BB will produce offspring all of this genotype; random mating would give a maximum of 50%.

Electrophoretically determined protein markers are of considerable value in a number of other aspects of animal and plant breeding. They can be used to check on breeding line purity, and to enable identification of progeny of several crosses kept in the same experimental unit (in order to minimize the contribution of environmental variability). Genetic marking is especially useful in breeding experiments involving fish and other aquatic animals as, unlike common farm livestocks, individuals cannot be artificially marked at birth. Electrophoretically determined protein variants may ultimately be instrumental in overcoming many of the difficulties which have resulted in fish farmers being '100 to several thousand years behind farmers of land animal species' (Moav *et al.*, 1976).

Intra-specific taxonomy of parasites

Differences in pathogenicity, susceptibility to chemical control, intermediate host and life-cycle characters among supposedly conspecific strains, have led to considerable interest in intra-specific taxonomy of parasites and their hosts.

The flagellate protozoans of the genus *Trypanosoma* are major disease-producing organisms affecting man and his livestock. Since trypanosomes are asexual organisms, each species is a genetically heterogeneous collection of clones. Isoelectric focusing in the pH range 3.5–9.5 and SDS electrophoresis have been used to characterize the strains of *Trypanosoma evansi* (Gibson *et al.*, 1978). In both methods the greatest similarities found were among several clones derived from one strain. It is of interest that, in this case, no differences were detected among these strains when horizontal starch gel electrophoresis followed by staining for 11 enzymes was used. However inter-strain differences have been detected by this latter technique in other *Trypanosoma* species (Bagster and Parr, 1973; Kilgour and Godfrey, 1973).

Trypanosoma cruzi, the causative agent of Chagas' disease, is a major cause of mortality on the South American continent. Isolates of *T. cruzi* from diverse mammals and insect vectors are morphologically indistinguishable. Using starch gel electrophoresis of six enzymes, Miles *et al.* (1978) found three genetically discrete forms of this species. Enzyme patterns were found to remain stable over a period of two years during numerous passages, in four different culture media. Enzyme patterns were also found to be independent of the host from which the organism was obtained. Strain characterization of *T. cruzi* has also been carried out using restriction endonuclease analysis of kinetoplast DNA (Mattei *et al.*, 1977). Isozymes of aspartate and alanine aminotransferases have been used to demonstrate the existence of three distinct types of *T. vivax*. It is a matter for debate whether these various strains of different *Trypanosoma* species should be regarded as separate species, as discussed for non-parasitic protozoa (p. 106).

Blood flukes of the genes *Schistosoma* are a further important pathogenic group. In the characterization of schistosomes, isoelectric focusing has been shown to be superior to other techniques (Ross, 1976). However, as yet definitive identification of all strains is not possible. Individual strains are preferentially transmitted by different snails, and this makes the correct classification of these intermediate hosts of particular importance. Thus taxonomy of snails of the genus *Bulinus* is of vital importance in the understanding of the epidemiology of human schistosomiasis. Problems in the taxonomy of this group have been outlined by Berrie (1973) and progress using biochemical techniques reviewed by Wright (1974).

Intra-specific systematics of man

Human intra-specific systematics has always been of special interest to biologists. Nei and Roychoudhury (1974) analysed allelic frequency variation in the three major races of man: Caucasoids, Negroids and Mongoloids. The genetic distances among these races were found to be similar to geographical populations of other organisms ($\bar{D} = 0.014$). They further found that inter-racial genetic variation in man is small (*c*.10%) compared with intra-racial variation. This is in sharp contrast to the conspicuous phenotypic differences observed in some morphological characters such as pigmentation and facial structure (Nei, 1975). However, it has been estimated that only three or four loci control the difference in pigmentation between Caucasoids and Negroids (Stern, 1970).

As with other organisms, there is heterogeneity among human populations in the occurrence and frequency of alleles at some loci. This variation has been used to trace the movements and origins of human populations. To date this has involved many blood group alleles (reviewed by Bodmer and Cavalli-Sforza, 1976) but a few protein loci are also suitable as markers. A variant allele at the superoxide dismutase locus (SOD^2) occurs in frequencies of less than 0.001 in most parts of the world, with the exception of Scandinavia, where it reaches 0.05 (Kirk, 1975). The SOD^2 allele is rare in most parts of Britain, but is found in appreciably higher frequency in the Orkney islands, presumably as a result of Viking influence in this area.

Sub-species and incipient species

As noted earlier, distinct sub-species show much greater genetic differentiation than populations which have not been deemed to merit sub-specific status. It may be appropriate at this point to reiterate the warning given in chapter 4 that, although conventionally electromorphs are regarded as the equivalent of genotypes, they may represent groups of genotypes with the same electrophoretic mobility in a given set of conditions. Hence genetic similarities given in table 5.1 are maximum values and may be substantially reduced as more detailed studies are performed.

A study of the xanthine dehydrogenase locus in *Drosophila pseudoob-*

Table 5.3 Allelic frequencies for 15 loci in samples of *Mus musculus musculus* and *M. m. domesticus* (see figure 5.6 for locations of sampling sites, based on Selander *et al.*, 1969).

Locus	*Allele*	*M. m. musculus samples*				*M. m. domesticus samples*	
		1	2	3	4	5	6
Esterase 1	a	1.00	1.00	1.00	0.95	–	–
	b	–	–	–	0.05	1.00	1.00
Esterase 2	b	–	0.17	0.07	0.05	1.00	1.00
	c	1.00	0.83	0.93	0.95	–	–
Esterase 3	a	0.30	0.67	0.47	0.32	0.40	0.30
	b	0.70	0.33	0.53	0.68	0.60	0.40
Esterase 5	a	1.00	0.86	1.00	0.95	0.78	0.69
	b	–	0.14	–	0.05	0.22	0.31
Alcohol dehydrogenase	a	0.83	0.70	0.90	0.60	0.97	0.97
	b	0.17	0.30	0.10	0.40	0.03	0.03
Cytosol NADP-malate	a	0.37	0.37	0.37	0.47	1.00	1.00
dehydrogenase	c	0.63	0.63	0.63	0.53	–	–
Mito. NAD-malate	a	–	0.03	0.03	–	–	–
dehydrogenase	b	0.95	0.93	0.80	0.71	1.00	1.00
	c	0.05	0.04	0.17	0.29	–	–
Hexose-6-phosphate	a	–	–	–	0.03	0.50	0.53
dehydrogenase	b	–	–	–	0.03	0.50	0.47
	c	1.00	1.00	1.00	0.94	–	–
6-phosphogluconate	a	0.33	0.83	0.93	0.76	1.00	1.00
dehydrogenase	b	0.67	0.17	0.07	0.24	–	–
NADP-isocitrate	a	0.17	0.23	–	0.08	0.93	0.93
dehydrogenase	b	0.83	0.77	1.00	0.92	0.07	0.07
Indophenol oxidase	a	0.02	0.13	–	0.42	1.00	1.00
	b	0.98	0.87	1.00	0.58	–	–
Phosphoglucomutase 1	a	0.75	0.43	0.40	0.37	1.00	1.00
	b	0.25	0.57	0.60	0.63	–	–
Phosphoglucomutase 2	a	0.03	–	0.23	0.13	1.00	1.00
	b	0.97	1.00	0.77	0.87	–	–
Phosphoglucose	a	1.00	1.00	1.00	1.00	0.93	0.93
isomerase	b	–	–	–	–	0.07	0.07
Haemoglobin	d	0.10	0.10	0.63	0.50	–	0.23
	s	0.90	0.90	0.37	0.50	1.00	0.77

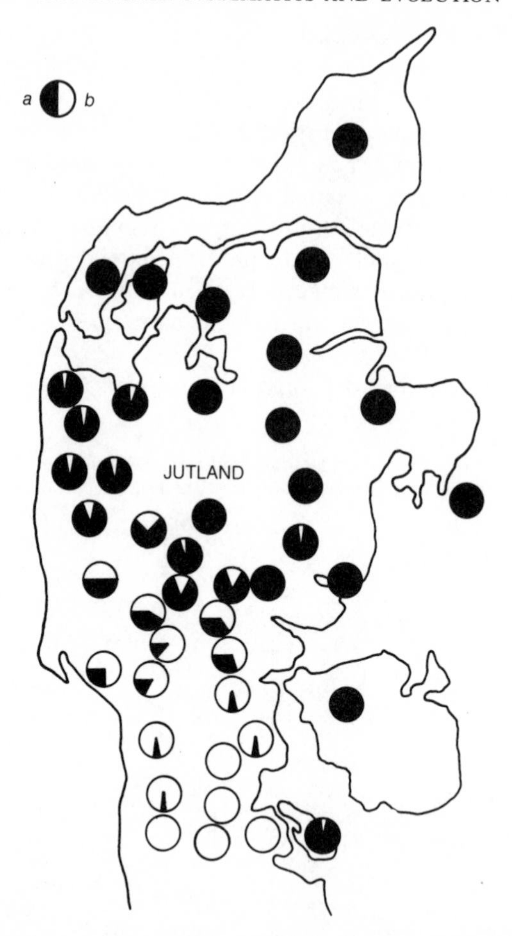

Figure 5.5. Geographical variation of the frequency of Esterase 1 alleles a and b in Danish populations of *Mus musculus*. Allelic frequencies in each sample are proportional to the area of the circle occupied by its symbol (redrawn from Hunt and Selander, 1973).

scura using four different electrophoretic conditions and a heat-sensitivity test, revealed 37 alleles where previously only six had been suspected. The isolated Bogotá sub-species *D. p. bogotana* was shown to be polymorphic for unique alleles at this locus, where formerly it had been thought to be monomorphic for the most common allele in the North American *D. p. pseudoobscura*. This is in keeping with the discovery that F_1 males obtained from mating of *D. p. bogotana* females with *D. p. pseudoobscura* males are sterile.

In some cases, populations ranked as sub-species may, in reality, merit species status, while in others, such as *D. pseudoobscura*, they may

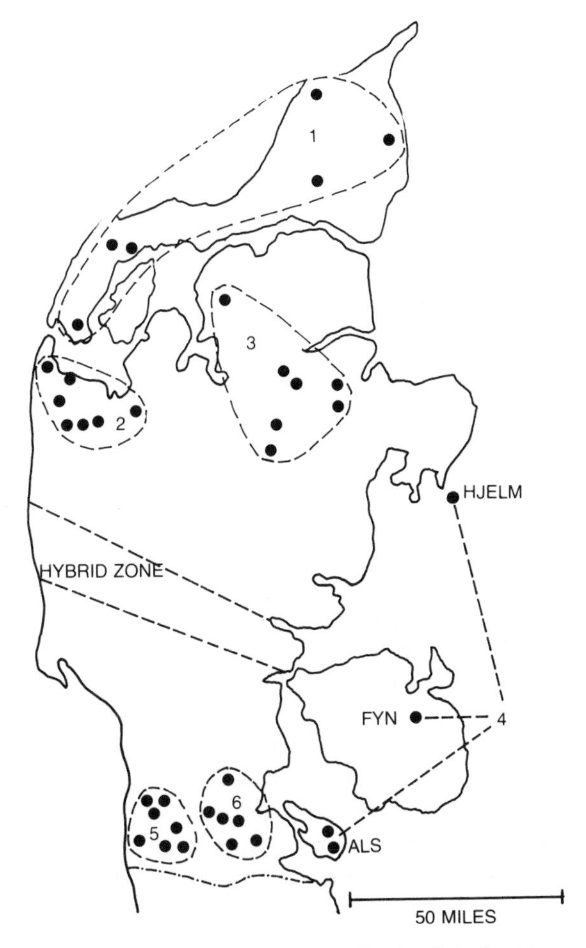

Figure 5.6. Map of part of Denmark showing sampling localities for *Mus musculus*. Allelic frequency data for polymorphic enzyme loci are given in table 5.3.

represent incipient species in the early stages of reproductive isolation. A further example of the latter situation concerns two sub-species of the house mouse, *Mus musculus musculus* and *Mus m. domesticus*, in northern and southern Denmark, examined by Selander *et al.* (1969). Twenty-five loci were identically monomorphic in the two sub-species. At 13 of the 17 polymorphic loci there were substantial differences in allelic frequencies between the sub-species. Different alleles were fixed in the two sub-species at two of these loci, and here heterozygotes were found only in a narrow zone between the ranges of the sub-species, indicating limited introgression (figures 5.5; 5.6; table 5.3).

CHAPTER SIX

SPECIES AND GENUS LEVEL SYSTEMATICS

AT THE LEVEL OF SPECIES COMPARISONS IT IS NECESSARY FOR MOST PURPOSES to take into account intra-specific variation due to environmental factors, age, polymorphism, etc. (chapter 4). Since this involves an examination of a number of individuals of each species, electrophoresis is the main method of analysis. Immunological and DNA-hybridization techniques, which are usually too insensitive for intra-specific studies, are used to a limited extent. For convenience, discussion of these latter techniques will be deferred to chapter 8.

Over the past two decades many electrophoretic comparisons of congeneric species have been carried out. These have involved from two to (in a minority of cases) ten or more species. The entire gamut of electrophoretic techniques and tissues have been used in these studies. Staining has been principally for general proteins and for a few enzymes such as esterases and lactate dehydrogenase. Less frequently twenty or more enzymes have been localized. The systematic content of many of these studies is low, frequently demonstrating only the morphologically obvious. Some studies are invalid because of failure to take account of the limitations of, or to control for artefacts involved in, electrophoresis.

In most cases where species have been examined for *sufficient proteins* and by means of high-resolution starch or acrylamide electrophoresis, or isoelectric focusing, *species-specific protein patterns* have been found. In general, the more closely related species are, the greater is the similarity in their electrophoretic patterns. These two observations form the basis of the use of electrophoresis in species- and genus-level systematics. First, protein characters can be used for the delimitation of species. Second, degrees of phenetic relationship can be determined and phylogenetic relationship inferred. For the latter purpose, a quantitative assessment of the similarities and differences between species is required. There are two principal types of electrophoretic approach.

Band-counting method

In this method the electrophoretic pattern is taken as an overall phenotype, no attempt being made to interpret the genetic basis of any intra-specific variability. In most cases, qualitative (mobility) differences only are used, although quantitative differences may be employed to aid in determining homologies between general protein bands. This approach is effective only with complex protein patterns derived from whole animal extracts, or tissues such as plasma. Separation is normally carried out by multiphasic acrylamide gel electrophoresis or isoelectric focusing, and the gels stained for general proteins or non-specific enzyme groups such as esterases. The complex electropherograms which result are likely to be similar only if they are derived from genetically similar organisms (figure 6.1).

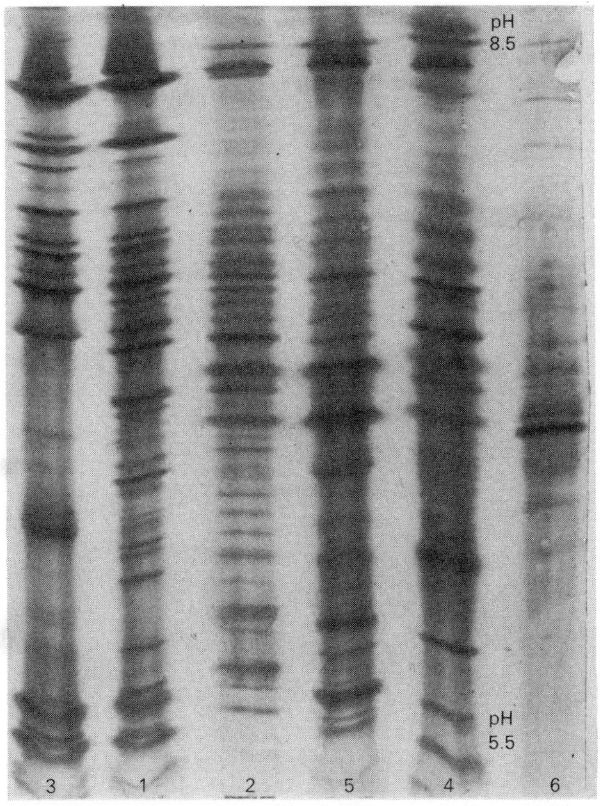

Figure 6.1. General protein patterns of white muscle extracts of several fishes. Separation of proteins was by isoelectric focusing in a pH gradient 5.5–8.5. Sample number 1, *Salmo trutta*; 2, *Salmo gairdneri*; 3, *Salmo salar*; 4, *Coregonus autumnalis*; 5, *Salvelinus alpinus*; 6, *Clupea harengus*. Note the species specificity of the patterns and also the greater similarity of the three *Salmo* species as compared with more distantly related ones.

Careful standardizations of sample extraction, gel preparation and electrophoretic conditions are required if inter-gel comparisons are to be valid. The similarity between pairs of patterns can be assessed in a number of ways. The most straightforward is a simple matching coefficient of similarity which is calculated as:

$$S_m = \left[\frac{\text{number of bands of common mobility}}{\text{maximum number of bands in an individual}} \right]$$

For example, the coefficients of similarity between the species shown in figure 6.2 are: $1:2 = 2/6 = 0.33$; $1:3 = 3/7 = 0.43$; $2:3 = 4/7 = 0.57$. Alternatively, a grid marked in millimetre or other suitable graduations may be superimposed on the gel, and the presence or absence of protein purified marker protein may be placed on the gel and the mobility of all protein bands in each sample calculated relative to this. These relative mobility values can be used to determine if homologous proteins on different gels are of the same mobility. Computations may also be carried out using the actual relative mobility values, where it is assumed that, for a given protein, greater mobility differences imply greater structural differences. Insufficient evidence is available to support or reject this premise at present. On samples separated by isoelectric focusing, the pI values can be used in a similar manner to relative mobility measurements.

Variation within species can be corrected for by carrying out multiple inter-specific comparisons and determining the variance of the similarity

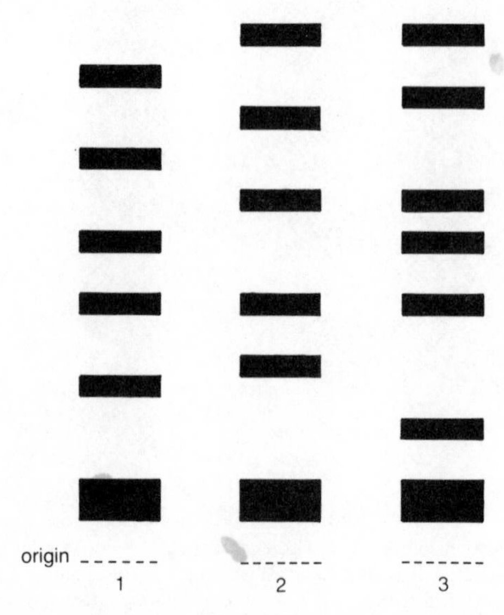

Figure 6.2. Hypothetical protein patterns of three species. See text for calculation of similarity coefficients.

coefficients. Alternatively, the frequency of occurrence of each protein band can be used as a character state. A sample prepared by pooling a number of individual extracts may dilute individual variable proteins below the level of detection.

In determining similarities between species, only mobilities of orthologous (p. 25) proteins should be compared. Electrophoretically similar (or coincident) general protein bands in two species may represent different proteins. Two or more 'superimposed' polymorphic proteins can give coincidental similarity in a general protein pattern. The only way to avoid this confusion is to stain for specific proteins, and to examine each one individually with regard to variation. Before discussing this individual protein approach, a consideration of the potential value of the general protein band-counting method is appropriate.

If a whole animal or complex tissue extract is subjected to polyacrylamide electrophoresis or isoelectric focusing, then some 20–40 protein bands will normally be resolved. Many of these bands will represent groups of proteins, but in a comparison of two congeneric species this effect should be common to both. If two species being compared share a common polymorphism, then two individuals may represent different homozygous states and give a spurious difference. Sarich (1977) has pointed out that the maximum probability for such an occurrence at a polymorphic locus with two alleles at a frequency of 0.5 is 1/8. In a complex pattern of 20–40 bands, and assuming that about one third of loci are polymorphic, then, at the most one or two bands might be incorrectly interpreted as being different. Even this low error level can be effectively eliminated by comparing two individuals per species–the 1/8 probability per locus is then reduced to 1/128. Little further information is obtainable by the use of a greater number of individuals, until sufficient are examined for a locus-by-locus type of analysis to be feasible. Much more valuable is an extension to several tissues and to very-high-resolution two-dimensional techniques.

Sarich (1977) has compared the results obtained from the study of plasma proteins and liver esterases in an individual from each of three species of kangaroo rats (*Dipodomys*) with the information obtained from a detailed study of 22 loci and 100 individuals. The latter approach had shown that *Dipodomys heermanni* is in reality two quite distinct species, with one of these more closely related to a third (*D. panamintinus*) than it is to the other *heermanni* species (now renamed *californicus*) (Patton *et al.*, 1976). Sarich found that with one gel and 'band-counting', congruent results were produced, although the D-values obtained were appreciably higher (table 6.1).

More insidiously hazardous for this approach, however, is variation due to other than polymorphism. Plasma protein patterns from conspecific individuals often show considerable variation in the number

Table 6.1 A comparison of the genetic distances among three *Dipodomys* species obtained with plasma proteins and liver esterases using one individual per population (above diagonal) with those obtained in a study of 22 loci in 100 individuals. See text for further explanation. Based on Sarich, 1977.

	D.h. (5)	*D.p.*	*D.h.* (4)
Dipodomys heermanni (5-toed)	0	0.47	1.60
Dipodomys panamintinus	0.08	0	1.35
Dipodomys heermanni (4-toed)*	0.24	0.15	0

* Now renamed *Dipodomys californicus.*

and mobilities of protein fractions (figure 6.3). In part this is due to the blood being the transport and drainage system of the organism, and thus varying in its composition with changing physiological state. This is particularly unfortunate, as plasma proteins are among the more rapidly evolving proteins, and thus suitable for use in systematic problems involving closely related species. Muscle extracts show much greater intra-specific uniformity, but these proteins appear to be conservative in their rates of evolution.

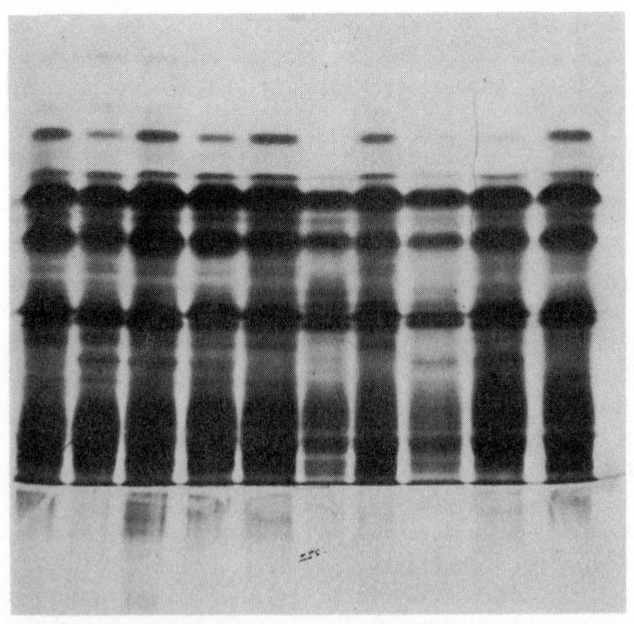

Figure 6.3. Electrophoregrams of plasma proteins of the brown trout *Salmo trutta*, produced by electrophoresis in polyacrylamide gel. The acrylamide concentration in the sample-concentrating portion of the gel was 5% and the buffer was Tris-HC1, pH 6.7. The separating gel had an acrylamide concentration of 7.5% in a Tris-HC1 buffer, pH 8.9. The electrode buffer was Tris-Glycine, pH 8.3. Note the complex individual variability in the patterns, which is very difficult to interpret in terms of polymorphism at individual loci.

In spite of the above limitations, the band-counting method, using high-resolution polyacrylamide electrophoresis of one or two individuals of each species, may be of value in giving a preliminary indication of whether or not electrophoresis will be useful in solving a particular taxonomic problem. Also, in situations where for practical or conservation reasons only one or two individuals are available, it may be the only feasible method. When taxonomic details have been elucidated, this approach may be employed in the routine identification of individuals. Coupled with image scanning and computer analysis, this identification method can become almost 'automatic'.

Locus-by-locus analysis

If staining is carried out for specific enzymes, their homologies among species are readily determinable. The bands representing each protein and their intra-specific variations can be analysed in genetic terms as discussed in chapter 4. Based on the allelic frequencies at each locus, Nei's coefficients of mean genetic identity ($\bar{I}$), mean genetic distance ($\bar{D}$), and time of divergence between each species pair can be calculated in the same way as outlined for populations (pp. 74, 166).

When inter-specific comparisons have been formulated in mathematical terms, either as 'band-counting' or Nei's coefficients, subsequent analysis can be carried out by the methods of numerical taxonomy, using computers to evaluate similarities, produce dendrograms and keys, and carry out identifications of unknown species.

Typical values of $\bar{I}$ and $\bar{D}$ for various congeneric species and confamilial genera are given in table 6.2. As might be expected, the genetic distance generally increases as the rank of the taxa being compared becomes higher. However, even species which are virtually identical in their morphological characters (sibling species) are shown to be considerably divergent at the structural gene level. On average, species show codon substitutions at about half of their loci.

Electrophoresis is most sensitive and reliable as a systematic tool in the range of one codon substitution per locus to about one codon substitution per ten loci. This corresponds to the range of differences found between, on the one hand, closely related genera and, on the other, divergent conspecific populations and closely related species. The distribution of loci of various degrees of genetic identity is shown in figures 6.4, 6.5 and 6.6 for sibling species, species and genera respectively. A very similar distribution for sibling species and species is found in both *Drosophila* and centrarchid fish, and may represent a common pattern for most outcrossing sexual organisms. The notable feature is that in all cases the distribution is essentially U-shaped, i.e. at a given locus, in most

Table 6.2 Mean values of genetic identity ($\bar{I}$) and genetic distance ($\bar{D}$) between taxa of various ranks.

Taxa	*Ref.*	*Sibling species*		*Distinct species*		*Genera*	
		$\bar{I}$	$\bar{D}$	$\bar{I}$	$\bar{D}$	$\bar{I}$	$\bar{D}$
Invertebrates							
Drosophila willistoni group	1	.563	.581	.352	1.056		
Drosophila obscura group	4			.300	1.204		
Drosophila mulleri group	1	.777	.252				
Hawaiian *Drosophila* group	1	.539	.618				
Asterias	1			.672	.397		
Pteraster–Diplopteraster	1					.257	1.359
Speyeria (butterflies)	6			.833	.182		
Vertebrates							
Californian minnows	3					.589	.529
Centrarchidae (sunfish)	2			.544	.626	.297	1.340
Taricha (salamanders)	1			.745	.296		
Taricha–Notophthalmus	1					.306	1.187
Anolis (lizards)	1			.667	.405		
Mus musculus–Mus spretus	5			.598	.514		
Plants							
Lupinus	1			.353	1.041		
Hymenopappus	1			.896	.109		
Clarkia	1			.280	1.273		
Mean of above taxa		.626	.484	.567	.645	.362	1.104

References:

1. Ayala, 1975
2. Avise and Smith, 1977.
3. Avise and Ayala, 1976
4. Marinković *et al.*, 1978.
5. Britton and Thaler, 1978.
6. Brittnacher *et al.*, 1978.

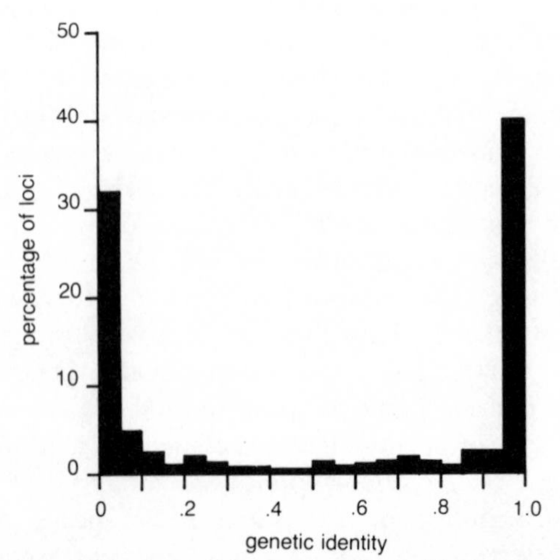

Figure 6.4. Histogram showing the percentage of loci within a given range of genetic identity among sibling species of the *Drosophila willistoni* group (redrawn from Ayala, 1975).

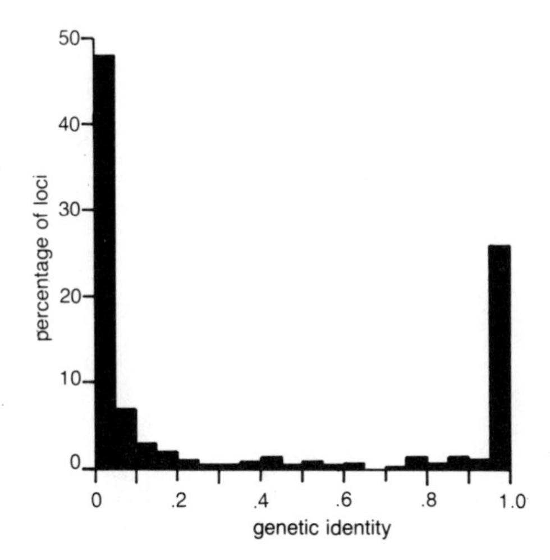

Figure 6.5. Histogram showing the percentage of loci within a given range of genetic identity among morphologically distinguishable species of the *Drosophila willistoni* group (redrawn from Ayala, 1975).

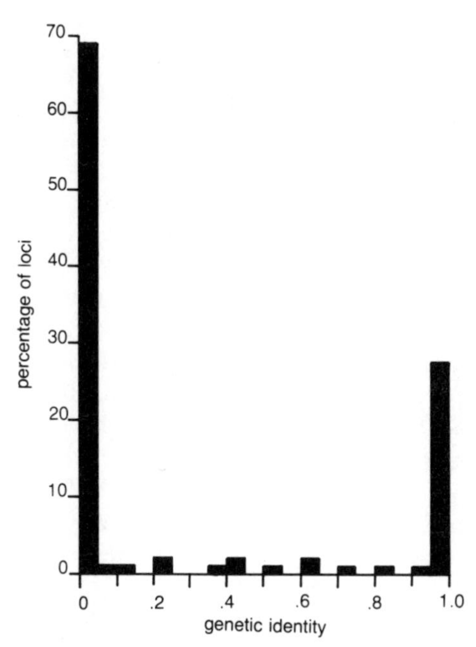

Figure 6.6. Histogram showing the percentage of loci within a given range of genetic identity among different genera of sunfish (redrawn from Avise and Smith, 1977).

cases, any two species are identical in allelic composition or else completely distinct with unique alleles. From a systematic point of view, this means that it is not essential to sample many individuals at each locus as, unlike inter-population comparisons (chapter 5), allelic frequency differences are not an important feature of inter-specific studies. The bimodal distribution of genetic similarity does, however, make it important that as many loci as possible be sampled. Screening of five individuals from each species for 50 loci will give much more useful systematic information than the examination of five loci in 50 individuals. At a minimum, 10 loci should be employed in inter-specific comparisons.

Dendrograms

When a number of species are compared for adequate loci, matrices of similarity and distance coefficients between all pairs of species may be formed (table 6.3). Although a large amount of information is contained in such a matrix, it is in a form which is difficult to visualize. For this reason a number of methods have been used to cluster taxa and produce dendrograms based on similarity and distance values. The three main methods used for analysing electrophoretic data are:

1. The unweighted pair group method with arithmetic means (UPGMA) (Sneath and Sokal, 1973).
2. The phylogenetic tree method of Fitch and Margoliash (1967).
3. The modified Wagner tree procedure of Farris (1972).

The UPGMA method is an agglomerative clustering procedure. The first cluster is formed from the most similar species pair, and this pair is henceforth treated as a single entity. The data matrix is recalculated, and a species which exhibits the highest level of similarity to a cluster is admitted to the previously formed clusters. Recalculation is continued in cyclical fashion until all species are clustered. In UPGMA, the average similarity of a potential member to a cluster is determined by weighting equally each species in the extant cluster (further details of UPGMA calculation are given in chapter 10). The resulting dendrogram is a useful visual summary of the relationships among species and groups of species (figure 6.7). Dendrograms so produced are, strictly speaking, phenograms and form evolutionary trees only if the rate of evolution is constant in different lineages for a given protein. The extent to which this may be the case is dicusssed in chapter 8. (It is not required that different proteins evolve at similar rates.)

The Fitch and Margoliash, and Farris procedures do not require homogeneous evolutionary rates and thus would seem to be more appropriate for the construction of phylogenies. Prager and Wilson

Table 6.3 Estimates of genetic identity (above diagonal) and genetic distance (below diagonal) among members of nine genera of Centrarchidae (sunfish) based on 11 loci (based on Avise and Smith, 1977).

	1	2	3	4	5	6	7	8	9	10
1. *Lepomis* (mean for 10 species)		0.323	0.308	0.484	0.202	0.207	0.192	0.288	0.304	0.330
2. *Pomoxis nigromaculatus*	1.142		0.365	0.374	0.109	0.091	0.183	0.185	0.241	0.275
3. *Enneacanthus obesus*	1.184	1.006		0.372	0.184	0.182	0.454	0.369	0.560	0.455
4. *Micropterus salmoides*	0.732	0.984	0.989		0.096	0.095	0.218	0.289	0.414	0.343
5. *Elassoma okefenokee*	1.634	2.214	1.691	2.341		0.829	0.260	0.281	0.198	0.369
6. *Elassoma evergladei*	1.610	2.393	1.705	2.355	0.187		0.182	0.277	0.195	0.364
7. *Centrarchus macropterus*	1.700	1.700	0.788	1.522	1.347	1.705		0.277	0.418	0.273
8. *Acantharchus pomotis*	1.322	1.685	0.997	1.242	1.271	1.285	1.285		0.290	0.370
9. *Ambloplites rupestris*	1.196	1.423	0.579	0.882	1.622	1.636	0.872	1.238		0.285
10. *Archoplites interruptus*	1.117	1.292	0.787	1.070	0.996	1.010	1.298	0.995	1.255	

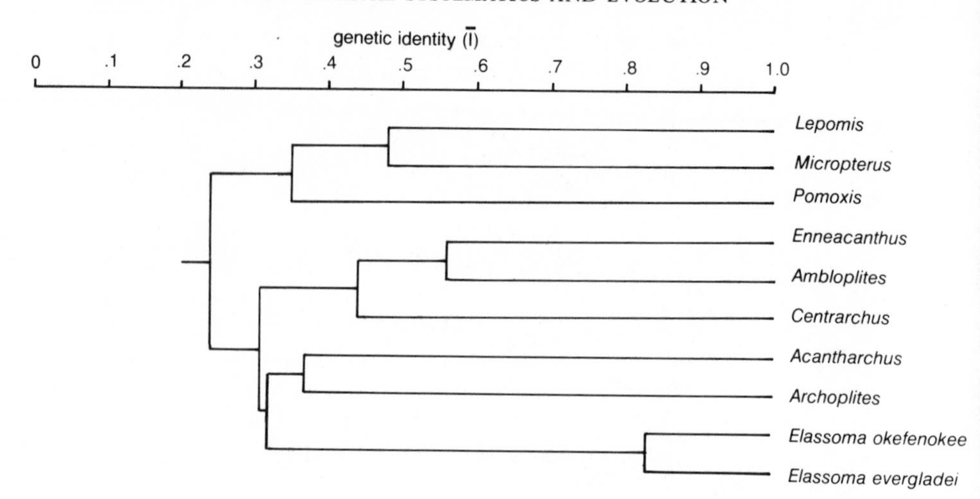

Figure 6.7. Phenogram showing relationships of nine genera of sunfish, generated according to UPGMA method of cluster analysis and based on the data in table 6.3.

(1978) compared the dendrograms resulting from the use of these three different methods on the same data sets. They examined the distortion which each method produced of the original data matrix and found that the Fitch-Margoliash procedure gave the best results for electrophoretic and micro-complement fixation data. The UPGMA procedure was on average better than the Farris method, but not as good as the Fitch-Margoliash procedure for analysing information derived from these techniques.

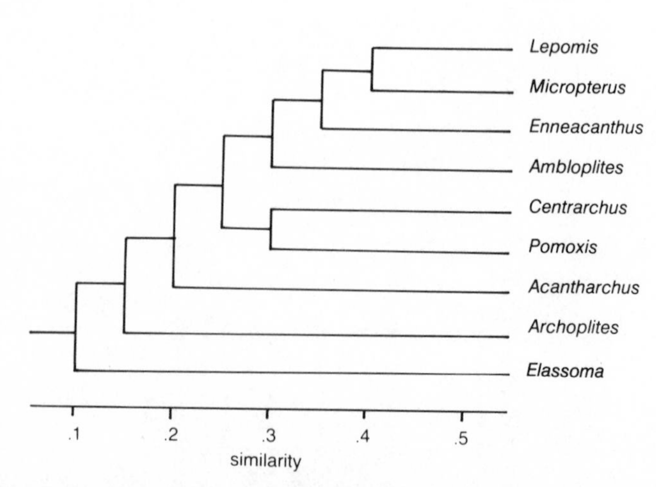

Figure 6.8. Dendrogram showing the relationships of nine genera of sunfish, based on a study of the acoustico-lateralis system. Compare with figure 6.7 (redrawn from Avise *et al.*, 1977).

Dendrograms constructed by the above methods from electrophoretic data can be compared directly with phenograms, cladograms, etc., produced from other information. In some cases, dendrograms constructed from electrophoretic studies of proteins show similar relationships with those produced on the basis of morphology (figure 6.8). Detailed studies of situations where protein relationships are in disagreement with those derived from other criteria may be rewarding in pinpointing cases of unusual speciation events or erroneous present classification.

The construction of an evolutionary tree is the first step in the study of the evolutionary processes involved in producing a group of related species. If rates of protein evolution are constant, protein differences would not only reflect phylogeny but would also indicate the actual times when the various cladogenic events took place, i.e. protein differences would serve as an evolutionary clock.

Evolutionary rates

Sarich (1977) has suggested that there are two main groups of electrophoretically studied proteins with respect to evolutionary rates. He points out that plasma proteins and some enzymes not involved in complex metabolic pathways appear to accumulate amino-acid substitutions some ten times more rapidly than do those enzymes normally sampled in electrophoretic surveys. This bimodality in the electrophoretically observed rates of protein evolution makes it incorrect to calculate a single genetic distance and relate this to a time scale. Sarich has calculated that the correct relationships are approximately:

1. $T\text{(years)} = 30 \times 10^6\, D$ for slowly evolving loci.
2. $T\text{(years)} = 2.4 \times 10^6\, D$ for rapidly evolving loci.

Nei's overall value was: $T = 5 \times 10^6\, D$ (p. 74) figure 6.9).

In the examination of recently diverged species (less than 5×10^6 years), plasma proteins and enzymes such as non-specific esterases, ribonuclease, lysozyme and carbonic anhydrase may be of more value than the glycolytic and other enzymes which principally have been used in electrophoretic studies.

Keys

Once a classification has been produced using electrophoretically determined protein characters, a dichotomous key can then be prepared for the identification of unknown species. An electrophoretic key can either compare protein mobilities with those of a control species (table 6.4) or use relative mobility or isoelectric point values.

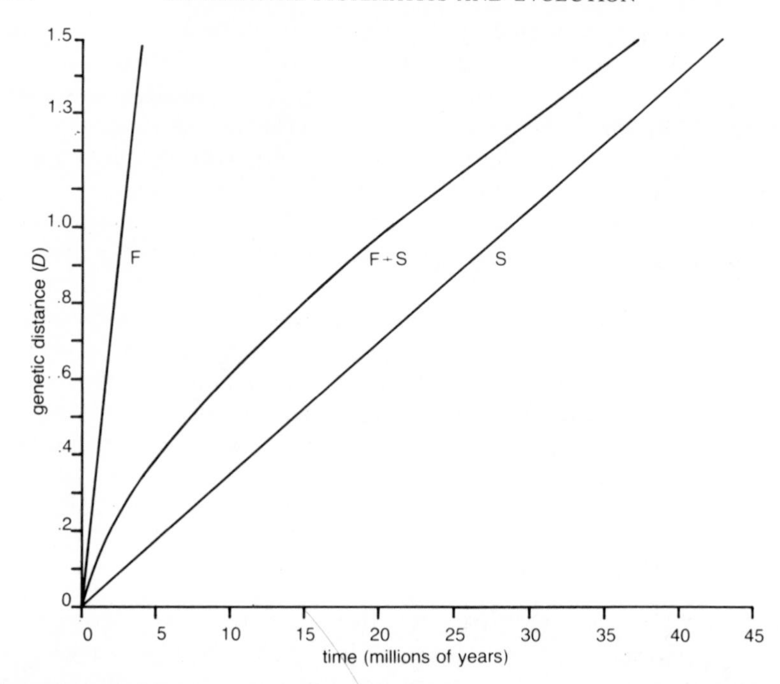

Figure 6.9. Rates of protein evolution. The rates of accumulation of electrophoretically determined genetic distance (D) for rapidly evolving proteins (F), slowly evolving proteins (S), and a mixed system involving $\frac{3}{4}$ slowly evolving, and $\frac{1}{4}$ rapidly evolving proteins ($F+S$) (redrawn from Sarich, 1977).

Table 6.4 Biochemical key to species of *Lepomis* (sunfish), using *Lepomis macrochirus macrochirus* as a standard, and taking mobilities of enzymes in other species relative to this (based on Avise, 1975).

A	Malate dehydrogenase-2; same mobility	B
	Malate dehydrogenase-2; faster in mobility	C
B	Glutamate oxalate transaminase-2; same mobility	*L. humilis*
	Glutamate oxalate transaminase-2; slower in mobility	*L. gulosus*
C	6-phosphogluconate dehydrogenase-1; same mobility	D
	6-phosphogluconate dehydrogenase-1; slower in mobility	F
D	Peptidase-2; slower in mobility	E
	Peptidase-2; faster in mobility	*L. gibbosus*
E	Indophenol oxidase-1; faster in mobility	*L. punctatus*
	Indophenol oxidase-1; same mobility	*L. auritus*
F	Glutamate oxalate transaminase-2; faster in mobility	*L. microlophus*
	Glutamate oxalate transaminase-2; slower in mobility	G
G	Peptidase-2; faster in mobility	*L. megalotis*
	Peptidase-2; slower in mobility	H
H	Indophenol oxidase-1; faster in mobility	*L. cyanellus*
	Indophenol oxidase-1; slower in mobility	*L. marginatus*

Examples of the application of electrophoresis to species problems

Undoubtedly the most important contribution of electrophoresis to species-level systematics has been in the provision of characters for the delimitation of sibling species and in the taxonomy of groups of 'lower organisms' such as bacteria, algae, fungi, protozoa and some parasites where in many cases the morphology is simple. Morphological similarity in these organisms often masks considerable genetic diversity. As can be seen from table 6.2, sibling species of *Drosophila* have electrophoretically detectable codon substitutions at approximately half of their loci. From the systematist's point of view, this means that an examination of a relatively small number of proteins will be sufficient for the characterization of such species. Admittedly *Drosophila* may be an exception, in view of Dobzhansky's suggestion that the evolution of external morphology in this genus has reached a high degree of perfection, and hence adaptive evolution proceeds largely by physiological changes.

The first example of the discovery of sibling species by electrophoresis was that of the holothurian *Thyonella gemmata* (Manwell and Baker, 1963). Two groups of this sea cucumber differ very slightly in their morphological appearance—one group being identified as 'stouts' and the other as 'thins'. Haemoglobin patterns were found to be consistently different for 'stouts' and 'thins'. Unique enzyme and general protein patterns were also given. Morphologically these two sibling species are extremely difficult to distinguish, even by examining the various spicules which are standard characters for discriminating between closely related holothurians.

As already mentioned (p. 12), the larvae of the midges *Chironomus cingulatus* and *C. plumosus* are inseparable on morphological grounds. However, haemoglobin and general protein electrophoregrams of larval haemolymph can be used to distinguish these and other chironomid species (figure 6.10).

The ability of electrophoretic methods to differentiate between similar species at juvenile stages has also been used for the separation of salmon *Salmo salar* and brown trout *Salmo trutta*, fry and parr, as well as for many other species which are not readily identifiable except as adults.

Four sympatric species of the *Drosophila virilis* complex have been shown to exist in central and northern Europe where, up to 1971, a single species only (*D. littoralis*) was described in this group (Lakovaara *et al.*, 1976). One of the four was discovered as a deviation from the Hardy-Weinberg expectations at several enzyme loci in supposed *D. littoralis*. Males of the four species have slight differences in their genitalia, but females are virtually indistinguishable on morphological criteria. Three nearly monomorphic species-specific enzymes provide good diagnostic characters for these species.

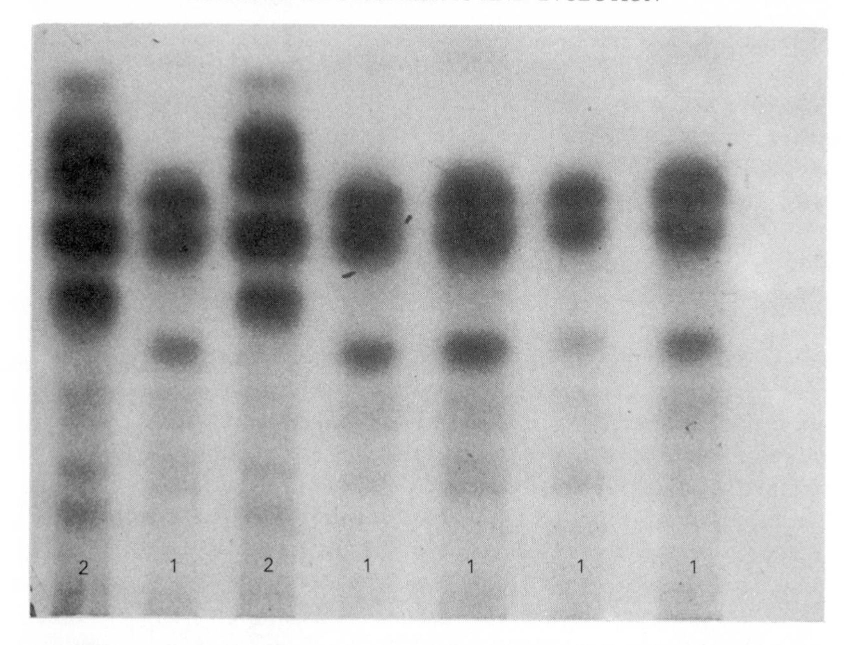

Figure 6.10. Electrophoretic characterization of sibling species of chironomids. The patterns were produced by horizontal starch gel electrophoresis of individual haemolymph samples, followed by staining with Amido Black. Pattern 1, *Chironomus cingulatus*; 2, *Chironomus plumosus*.

Sibling species complexes have been described for several ciliate protozoans. In *Tetrahymena pyriformis* there are thirteen non-interbreeding groups or syngens. Borden et al. (1977) examined 12 syngens and found that each gave unique zymograms. The most closely related syngens were unique in about two-thirds of their enzyme fractions. Several polymorphic proteins were noted in one syngen where 55 strains were examined. However, in this case the mobility changes due to polymorphisms were much smaller than the inter-syngen differences. The differences among these syngens are such as to suggest, using Nei's formula, that they have been evolving along separate lines for almost one million years. Although *Tetrahymena* syngens are morphologically indistinguishable, their reproductive isolation and degree of genetic differentiation argue for their establishment as separate species. Now that their characterization can be based on relatively quick and simple electrophoretic methods and without the need for living reference strains, Borden *et al.* contend that 'the chief impediment to their official designation as named species is removed'.

Paramecium aurelia is also a complex of at least 14 syngens, and here again enzyme electrophoresis shows a high genetic distance between all but one pair of syngens (Tait, 1970; Adams and Allen, 1975; Allen and

Gibson, 1975). Electrophoresis is thus a valuable method in the systematics of protozoans.

General protein patterns produced by isoelectric focusing of whole-animal extracts have been applied to the taxonomy of tapeworms of the troublesome *Diphyllobothrium* group (Bylund and Djupsund, 1977). Except in the first larval stage, these worms lack hooks, mouth-parts, etc., which in many parasites are the main morphological taxonomic characters. Another difficulty in the taxonomy of this group which is overcome by the use of proteins, is the extreme range of morphological variability which can be induced by different host species and varying physiological conditions.

Coluzzi and Bullini (1971) have demonstrated that there are two sibling species within the *mariae* complex of the *Aedes* mosquitoes. These exist as two allopatric forms: *zammitii* of the eastern and *mariae* of the western Mediterranean. They have slight morphological differences which are statistically significant between populations as a whole, but do not permit a reliable identification of individuals, especially in cases where hybrids may be present. The two groups have been shown to be monomorphic for different phosphoglucomutase alleles, and artificially produced hybrids exhibited a heterozygote pattern. Coluzzi and Bullini carried out a massive release experiment in which 25 000 larvae and pupae of the *zammitii* type were introduced into a *mariae* population in a region of restricted suitable habitat and after equivalent reduction of the *mariae* population. Subsequent sampling of over 2000 larvae and pupae in the area showed that less than 0.4% had a hybrid enzyme pattern. It seems likely, then, that efficient pre-mating isolating mechanisms exist between *zammitii* and *mariae*, and therefore these two forms should be regarded as separate species. In laboratory cages the two species interbreed freely. This emphasizes the need for caution in extrapolating from such laboratory experiments to the natural situation (p. 2).

Mediterranean house mice are of two types: those that breed permanently outdoors and those that breed indoors. 'Outdoor' mice are frequently shorter tailed than 'indoor' mice. Britton and Thaler (1978) have shown that these two morphotypes are homozygous for different alleles at four loci, even in mice captured a few feet apart. In view of this reproductive isolation they suggest that these mice should be referred to two different species, *Mus musculus* and *M. spretus*, in contradiction to the present conspecific status used by most workers.

The role of the most recent Pleistocene glaciations in producing sibling forms of salmonid fishes has already been discussed (pp. 83–85). The white fish (*Coregonus* species) comprise one such group and are found throughout the cooler regions of the northern hemisphere. They exhibit extreme phenotypic plasticity, which complicates the use of morphological characters in the study of their evolution and systematics. Most morphological characters, with the exception of the number of gill rakers, have been shown to be environmentally labile. As a group, the coregonines present a challenge to the biochemical systematist to unravel the mosaic of variable forms.

In a number of lakes in Britain and Ireland coregonine fishes are found. Early biologists referred these populations to three species. Later workers classified the Irish *Coregonus pollan* with one of the two British species, *C. lavaretus* or *C. albula*, although they were divided in their opinions as to which it was conspecific. This uncertainty arose from the intermediate nature of the morphology of the Irish fish when compared with the British ones. Electrophoretic studies of general proteins and enzymes (Ferguson, 1974; Ferguson *et al.*, 1978) have shown that the Irish *C. pollan* is in fact conspecific with the Alaskan-Siberian species *C. autumnalis* (figures 6.11, 6.12, 6.13 & 6.14). As such, it is the only

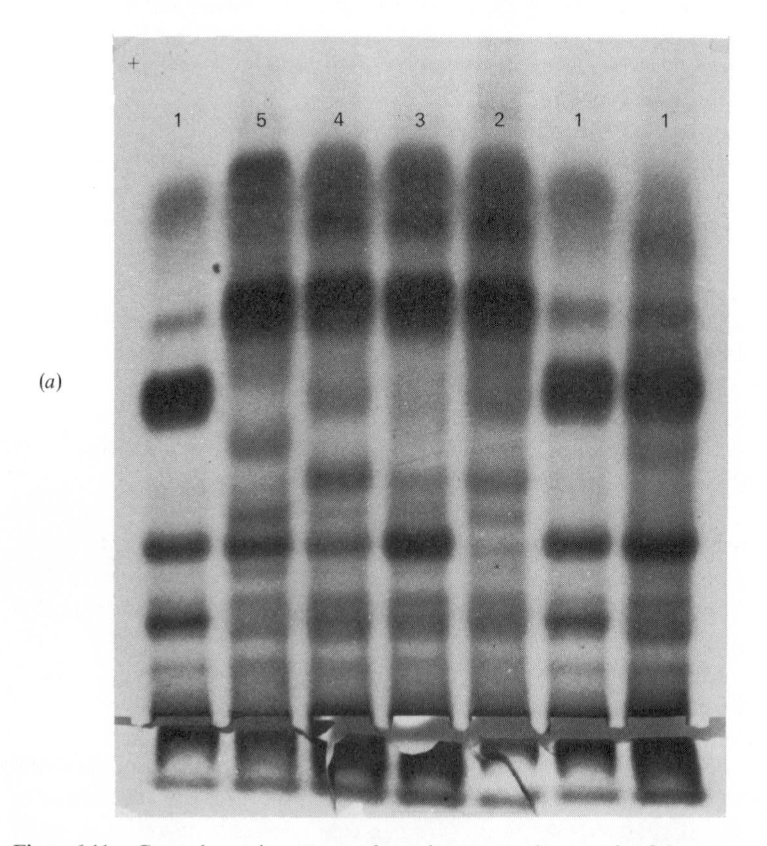

Figure 6.11. General protein patterns of muscle extracts of coregonine fishes.
(*a*) As separated by vertical starch gel electrophoresis using the Scopes (1968) buffer system. Sample 1, *Coregonus autumnalis pollan*; 2, *C. oxyrhynchus* (gwyniad); 3, *C. lavaretus* (powan); 4, *C. oxyrhynchus* (schelly); 5, *C. albula*.

(*b*)

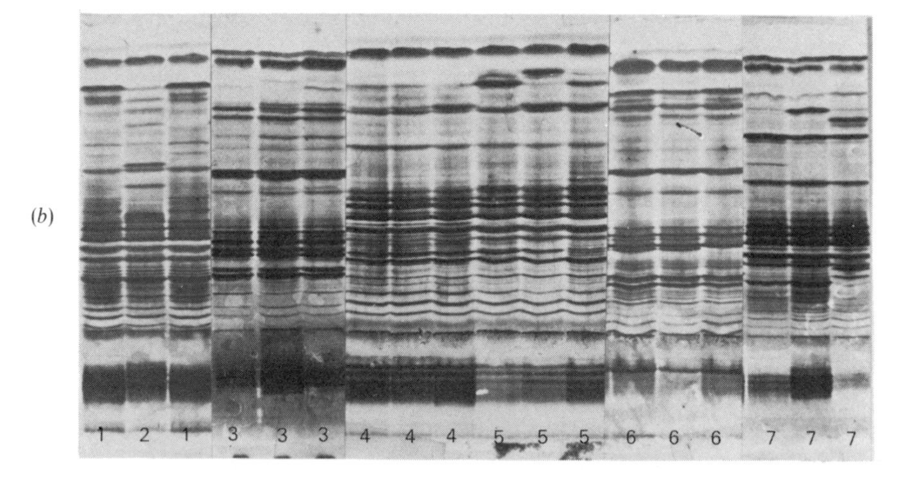

Figure 6.11. contd.

(*b*) As separated by isoelectric focusing in a pH gradient, 3.5–9.5 (highest pH at top of photograph). Sample 1, *Coregonus albula*; 2, *C. albula* × *C. lavaretus* hybrid; 3, *C. autumnalis pollan*; 4, *C. lavaretus*; 5, *C. nasus*; 6, *C. peled*; 7, *C. pidschian*. Courtesy of K-J. M. Himberg, Åbo Akademi.

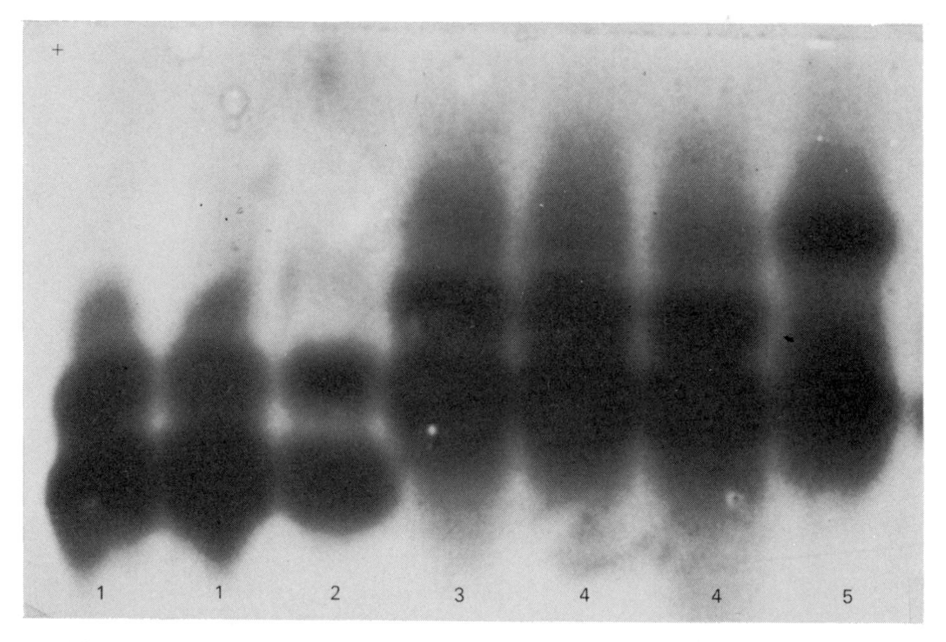

Figure 6.12. Creatine kinase zymograms of white muscle extracts of coregonine fishes as separated by vertical starch gel electrophoresis at pH 8.6. Sample number 1, *Coregonus autumnalis pollan* (Ireland); 2, *C. autumnalis* (Alaska); 3, *C. lavaretus*; 4, *C. oxyrhynchus*; 5, *C. albula*.

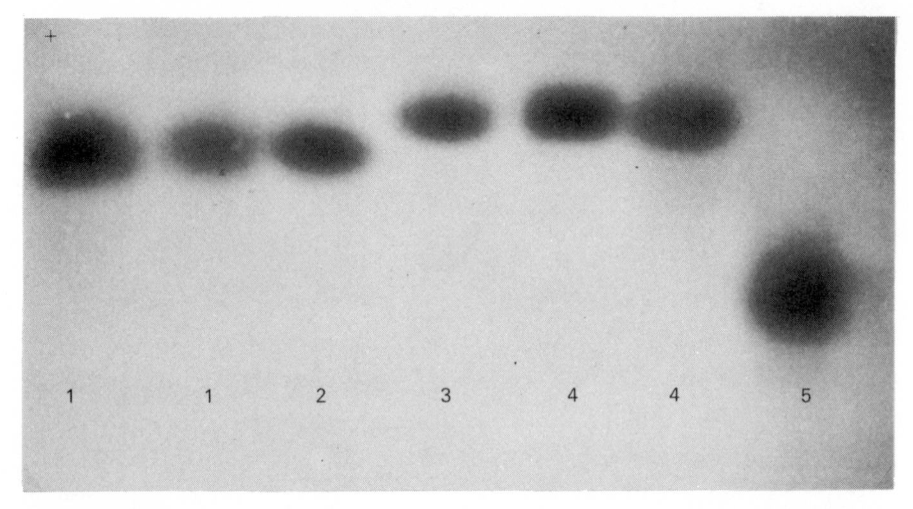

Figure 6.13. Valyl-leucine peptidase zymograms of white muscle extracts of coregonine fishes as separated by vertical starch gel electrophoresis at pH 8.6. Sample number 1, *Coregonus autumnalis pollan* (Ireland); 2, *C. autumnalis* (Alaska); 3, *C. lavaretus*; 4, *C. oxyrhynchus*; 5, *C. albula*.

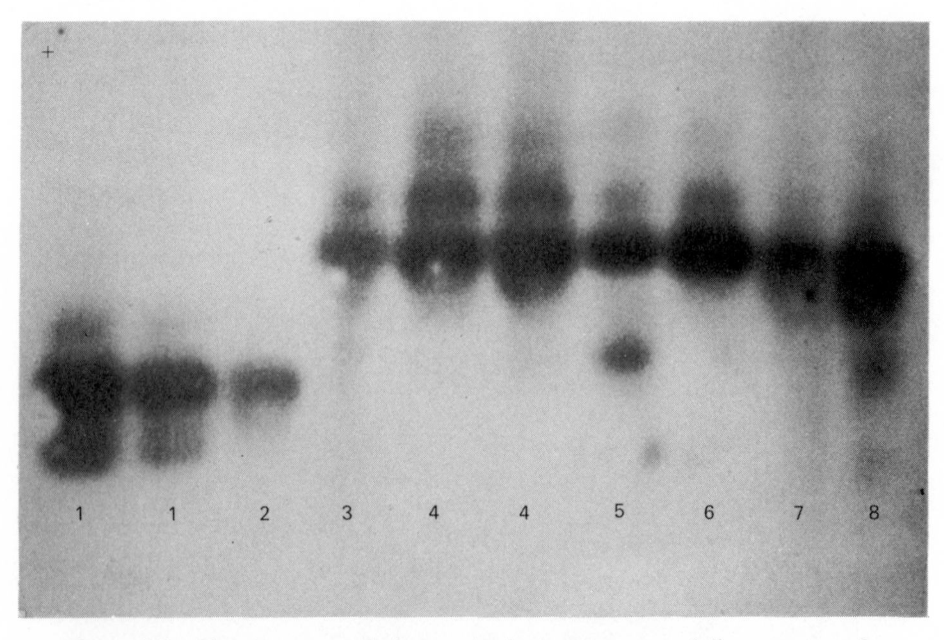

Figure 6.14. Phosphopyruvate hydratase (enolase) zymograms of liver extracts of coregonine fishes as separated by vertical starch gel electrophoresis at pH 8.6. Sample 1, *Coregonus autumnalis pollan* (Ireland); 2, *C. autumnalis* (Alaska); 3, *C. lavaretus*; 4, *C. oxyrhynchus*; 5, *C. albula*; 6, *C. pidschian*; 7, *C. muksun*; 8, *C. peled.*

representative of this species in western Europe. This relict distribution can probably be explained on the basis of events associated with the last Ice Age.

Forensic applications

Characterization of otherwise distinct species by electrophoresis is useful when only part of an organism is available. This approach has been used in the identification of blood and other tissues in forensic medicine, in identification of 'meat' confiscated from poachers and in cases of 'false description' in trading.

Genus and higher-category systematics

With the exception of closely related genera, electrophoresis is of limited systematic value at the genus and higher-category level. In this range, generally more than one electrophoretically detectable codon substitution per locus has taken place. This leads to the possibility of charged substitutions cancelling each other. Due also to the finite number of positions which a protein can occupy on the electrophoretic gel, the chance of two very different proteins having coincident mobilities becomes much higher.

There are certain exceptions to this generalization. Some proteins evolve at much slower rates than others, and electrophoretic comparisons of these may be of value at the intra-familial level. Structural proteins, perhaps because of their involvement in cell architecture, and proteins such as those of egg white of birds appear to have evolved at slow rates. When the overall electrophoretic pattern (rather than individual proteins) is used for comparison, then the probability of coincidental similarities is minimized.

Avian egg-white

Sibley and his associates (1960; 1970; 1972*a*; 1972*b*; 1976) have carried out extensive comparisons of the egg white of some 1500 species of birds. Initially, paper electrophoresis was employed in these studies, then vertical starch gel, and in later years the superior resolution of isoelectric focusing was utilized. These comparisons enable a number of rearrangements of avian higher categories (graded into, 'highly probable, probable, possible and improbable' judgments) to be made about bird relationships. Among their many findings were the monophyletic nature of the large ratites; the close relationship of the hoatzin *Opisthocomus hoazin* to cuckoos of the subfamily Crotophaginae: and the greater similarity of the flamingos to the Ciconiiformes (herons and storks) than to the Anseriformes (ducks and geese). As well as producing information of value in avian systematics, these studies provided impetus to biologists working with other organisms, because of the way in which electrophoretic evidence was incorporated with that derived from other taxonomic characters and the critical approach taken to the significance of the findings.

Feathers

Brush (1976) has applied polyacrylamide gel electrophoresis to the examination of bird feather proteins, and has found that information derived from these electrophoregrams is in close agreement with the accepted taxonomy of the Anseriformes (ducks and geese) at the tribe and sub-family levels.

SDS electrophoresis

In a novel approach, Snyder (1977) has examined soluble proteins from 49 species of bees (Apoidea) using SDS electrophoresis. Since this method separates denatured polypeptides solely on the basis of their molecular weights, it produces a very conservative set of characters which may be applicable at a higher taxonomic level than conventional electrophoretic treatment. Thus individual amino-acid changes which may alter the charge will have little or no effect on the molecular weight of a polypeptide. Only when substantial amino-acid changes take place will the molecular weight, and hence the mobility under SDS electrophoresis, be changed.

The bees are well studied morphologically and thus make a very suitable group to test the application of the SDS technique to systematics. As might be expected with conservative characters, individual variation was found to be insignificant. Snyder tried two different approaches to the analysis of the resulting electrophoretic patterns. One was based on molecular-weight (mobility) comparisons and the other on relative absorption values (quantity) of equivalent proteins as measured by spectrophotometric scanning. These two methods resulted in somewhat different results, and he indicated that the latter method is more valuable. Using comparisons of absorption values, Snyder found that some morphologically similar groups were biochemically differentiated and *vice versa*.

The genetic basis of absorption values is questionable, however. Absorption changes could be due to changes in regulatory genes altering the quantity of a particular polypeptide. Concentration differences might also be due to differences in the amount of stain binding as a result of structural changes in the polypeptide. This latter possibility should be easily testable with proteins of known primary structure. Homologous enzyme bands in closely related species can stain different colours, even when on the same gel, e.g. roach and rudd esterase (p. 115). It is probable, of course, that each general protein band in the SDS technique is a group of several different proteins, and the composition as such may vary among distantly related taxa.

Duplicate loci

The detection of duplicate loci by electrophoresis may have some applications in higher-category systematics. The large number of electrophoretically complex enzyme patterns was one of the clues which led to the discovery that salmonid fish are tetraploid. All fish of the super-order Teleostei have at least two loci for phosphoglucose isomerase. More-ancient actinopterygian fish of the super-orders Holostei and Chondrostei have only a single PGI locus. Thus the duplication must have occurred after the lineage leading to the modern teleosts diverged from the holostean ancestor. This suggests a monophyletic origin for the teleosts.

CHAPTER SEVEN

HYBRIDS AND POLYPLOIDS

HYBRIDS HAVE ALWAYS BEEN OF INTENSE INTEREST TO THE BIOLOGIST, particularly as they are sometimes seen as antithetical to the biological species concept of reproductively isolated entities. In most cases, hybrids are morphologically intermediate between the parental species, and there is no difficulty in recognizing them as such. However, when hybrids are formed between similar or sibling species, where even separation of the parental species may be difficult, then identification of the hybrid can present problems to the systematist. In a minority of situations the hybrid may be morphologically indistinguishable from one of the parents.

Electrophoretic characterization of hybrids

Since even sibling species differ considerably in their structural genes (table 6.2) and since proteins are normally co-dominantly expressed, electrophoresis provides a valuable tool for the characterization of hybrids. In many cases the hybrid shows a summation of the parental electrophoretic patterns. Where different alleles are present at a locus in the parents, the hybrid will exhibit a 'heterozygote' type of pattern. Thus, if two species exhibit bands of different electrophoretic mobility for a particular protein, then the hybrid will show both bands, each at approximately half the concentration found in the parent. In the case of multimeric proteins, unique heteromers may be shown in the hybrid pattern, further facilitating identification.

Fish hybrids

The use of electrophoresis in the characterization of F_1 hybrids can be exemplified by the following study.

Roach *Rutilus rutilus* and rudd *Scardinius erythrophthalmus* are morphologically similar cyprinid fishes which occur sympatrically throughout Europe. Due to overlap in conventional fish taxonomic characters, it is difficult to distinguish roach and rudd from some localities, and even more so their hybrids. Difficulties in the identification of cyprinid hybrids have been recognized for a long time. Roach × bream hybrids were originally thought to be a separate species (Pomeranian bream). The British Record (Rod Caught) Fish Committee requires that all potential record fish of a species known to hybridize with a larger species, be examined by an expert. Several potential records have been invalidated due to their being probable hybrids.

Based on zymograms of esterases and lactate dehydrogenases, Brassington and Ferguson (1976) developed methods for the definitive identification of roach × rudd and other cyprinid hybrids. Roach and rudd each show a unique esterase pattern consisting of one major and three minor zones of activity. Roach also show a di-allelic polymorphism of the major fraction. Esterase zymograms with bands of the same mobilities, though differing in the intensities of staining of individual fractions, are given by extracts of eye, heart muscle, skeletal muscle and whole 2-cm standard-length fry. The major esterase isozyme has a greater mobility in rudd than in roach. In rudd-roach hybrids, two bands of activity are present, corresponding to the expression of the rudd fraction and one of the two roach homozygote bands (figure 7.1). There is also a colour difference between the two species in the staining of this isozyme, when 1-naphthyl acetate and Fast Blue RR are used for enzyme localization.

In roach and rudd, six lactate dehydrogenase (LDH) isozymes are found on staining the electrophoretic gel. These isozymes correspond to the five A-B tetramer series and the C^4 homotetramer (p. 48). The A^4 and C^4 isozymes have the same mobility in both species, but the B^4 isozymes, and other isozymes containing the B sub-units, have a greater mobility in the rudd than in roach. In the hybrids, four types of LDH polypeptide appear to be produced: A, rudd B, roach B, and C. Correspondingly the hybrids should theoretically show 16 LDH isozymes (C^4 plus 15 tetramer combinations of A, rudd B, and roach B). In fact, due to coincident electrophoretic mobility of different sub-unit combinations, only 11 bands of LDH activity are observed (figure 7.2) in the hybrid. The pattern, however, is clearly distinct from the parental six-banded zymogram and enables definitive hybrid identification.

Child and Solomon (1977) have shown that esterase and LDH are also expressed in rudd and roach serum, and also that serum general proteins and phosphoglucose isomerase can be used for the identification of roach-rudd hybrids.

Electrophoresis of serum is likewise valuable for the characterization of hybrids between other similar fish species. Payne *et al.* (1972) and

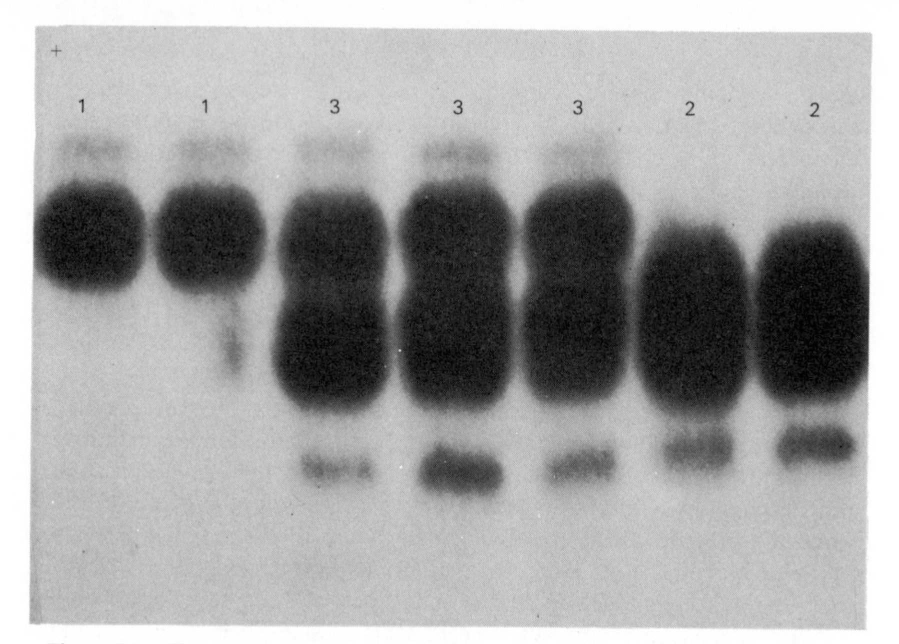

Figure 7.1. The use of esterase zymograms for the identification of hybrids between cyprinid fishes. Patterns were produced by vertical starch gel electrophoresis of eye extracts. Sample 1, rudd *Scardinius erythrophthalmus*; 2, roach *Rutilus rutilus*; 3, rudd × roach hybrid.

(*a*)

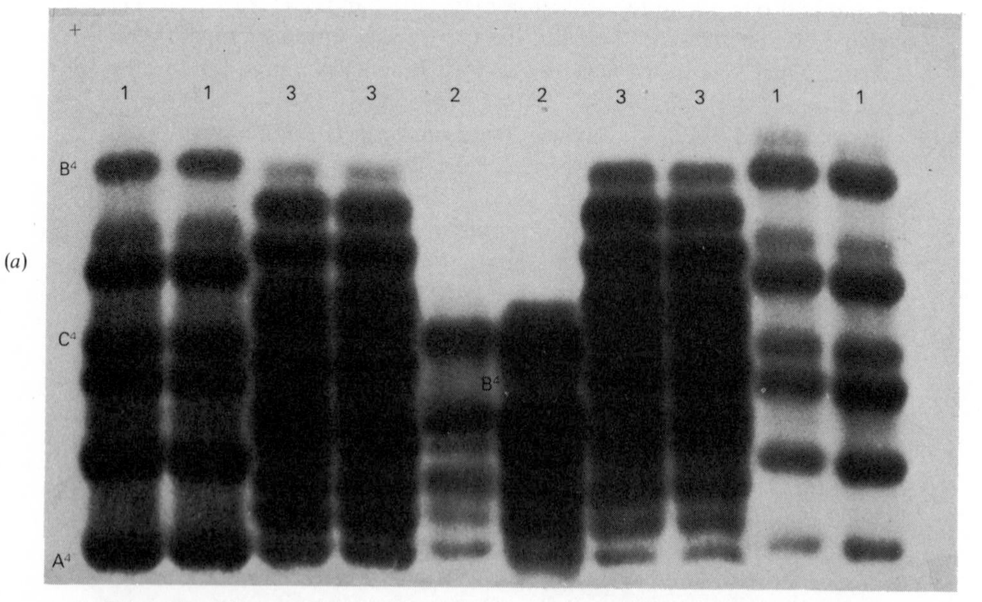

Figure 7.2. The use of lactate dehydrogenase zymograms for the identification of hybrids between cyprinid fishes.

(*a*) Patterns were produced by vertical starch gel electrophoresis of eye extracts in a pH 8.6 buffer system. Sample 1, rudd, *Scardinius erythrophthalmus*; 2, roach, *Rutilus rutilus*; 3, rudd × roach hybrid. The A^4, B^4 and C^4 homotetramers are labelled.

(*b*)

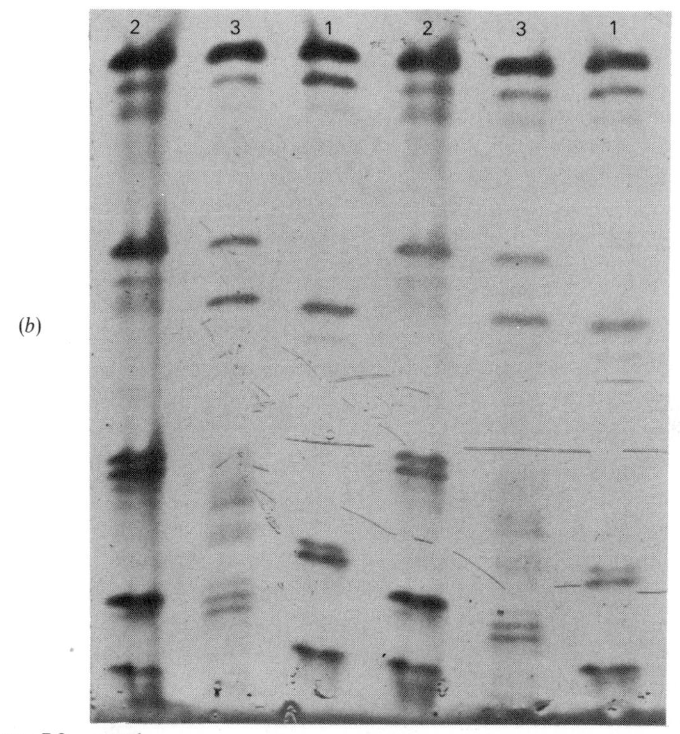

Figure 7.2. contd.
(*b*) Patterns were produced by isoelectric focusing of eye extracts in gels of pH range 5.5–8.5 (highest pH at top of photograph). Sample identification as in (*a*).

Solomon and Child (1978), using electrophoretic patterns of serum proteins, found that 28 out of 9166 fish (0.3%) caught around the British coast and morphologically identified as salmon were in fact salmon × trout hybrids.

Summations of parental general protein and enzyme electrophoretic patterns have been observed for a wide range of animal and plant F_1 hybrids.

It is to be expected in F_2 and later filial generation hybrids, and in backcross hybrids, that due to the independent segregation of many alleles, some loci should exhibit parental patterns and others hybrid combinations. However, Nyman (1970) showed that in F_2 salmon × trout hybrids, the trout alleles appear to be 'dominant'. This resulted in the salmon proteins not being expressed, although a few bands not present in either of the parents were observed. In a later study (Nygren *et al.*, 1975) it was shown that backcross hybrids (F_1 × salmon) exhibit both hybrid and salmon protein components. In other plant and animal F_2 hybrids, the expected combination of parental and hybrid proteins has been found.

Allelic repression and asynchrony

At certain enzyme loci in some F_1 hybrids, a summation of the parental patterns is not observable due to repression of one or, rarely, both alleles. Also, in some cases, one allele may show delayed activation. In most instances it is the paternal allele which shows repression or delayed activation. Allelic repression and asynchrony are much more common in hybrids derived from distantly related species than from closely related ones (Whitt *et al.*, 1973, 1977). The extent of allelic repression and asynchrony would seem therefore to be a possible method of indicating inter-specific differences in regulatory genes. In no case of a viable hybrid have more than about 40% of loci demonstrated repression (table 7.1) and so even in these cases there are still many suitable characters available for hybrid identification.

Lucotte *et al.* (1978) have noted summation of the parental electrophoretic patterns of albumin, immunoglobulin and haemoglobins in chicken × quail hybrids. The patterns, however, were asymmetrical with predominance of one or the other parental form. In the case of albumin and immunoglobulin, the quail fraction predominated, while for haemoglobins it was the chicken components.

In female mammals, dosage compensation of genes located on the X chromosome is produced by inactivation of one of these chromosomes (Lyon hypothesis). Since either the paternally or maternally derived X chromosome may be inactivated, this results in hybrids (and intra-specific heterozygotes) showing either the maternal or paternal form of sex-linked enzymes, depending on which X chromosome is inactivated, e.g. in female hybrids between the Arctic fox *Alopex lagopus* and silver fox *Vulpes vulpes*, some cells show the Arctic fox glucose-6-phosphate

Table 7.1 Allelic repression and genetic distance among sunfish (family Centrarchidae) and among macaque monkeys (based on Avise and Duvall, 1977).

Hybridizing species or subspecies	*Number of loci examined for allelic repression*	*% of loci exhibiting allelic repression*	*Genetic distance*	*Total number of loci examined*
Sunfish				
Lepomis macrochirus macrochirus × *L. macrochirus purpurescens*	15	0.0	0.181	15
Lepomis macrochirus × *L. microlophus*	6	0.0	0.948	11
Lepomis cyanellus × *Micropterus salmoides*	14	21.4	0.769	11
Lepomis microlophus × *Chaenobryttus gulosus*	5	40.0	0.980	14
Macaques				
Various pairs of 6 species	21	0.0	0.164	21

dehydrogenase pattern while others exhibit a silver fox zymogram (Serov *et al.*, 1978). Males, of course, show only the maternal form of the enzyme. Unlike the situation in the fox where either X chromosome may be inactivated, in horse *Equus caballus* × donkey *E. asinus*, hybrids (mule, hinny), it is the maternal X chromosome which is inactivated in all tissues, as judged from the glucose-6-phosphate dehydrogenase zymogram (Hutchison *et al.*, 1974).

Evolutionary significance of hybridization

The evolutionary significance of hybridization is a topic about which there has been much discussion but on which, in animals at least, there is little direct evidence. Hybrids formed between two species will be heterozygotes at corresponding loci at which the parental species are homozygous for different alleles. Since genetic variability is the essence of adaptation, this increased heterozygosity may allow rapid evolution, especially in conditions of varying environments as a result of geological or human disturbances. In this situation the hybrids may form a new 'superior' species, with the consequent elimination of the parental types. Alternatively, the hybrids may co-exist with the parental species, perhaps in an ecologically intermediate area, and may eventually form a third species. Such hybrids may be restricted to a hybrid zone, i.e. an area where only hybrids are found. This is possibly the situation shown by the crows *Corvus cornix* and *C. corone* (p. 5). Short (1973) has estimated that hybridization has been a significant factor in the recent evolution of at least 15% of the Nearctic bird fauna.

Backcrossing of the hybrid with one of the parental species may result in the introduction of genes from one species to the other. This is called *introgressive hybridization* or *introgression* and can increase the genetic variability within that species, thereby supplementing that produced by mutation. This may have the advantage that the blocks of introduced genes have already proved their value in another species in which they have been built up over several thousand years.

A possible example of advantageous introgression was reported by Lewontin and Birch (1966). The fruit fly *Dacus tryoni* has dramatically extended its range in Australia during the past 100 years. This extension appears not to have been due to increased food resources but by adaptation to extreme temperatures. The rapid genetic change necessary for the sudden evolution of this ecological tolerance seems to have been produced by introgression of genes from another species *D. neohumeralis*. Lewontin and Birch strengthened this suggestion by laboratory experiments which showed that, although the early generation hybrids were less well adapted than the parental species, as the hybrid population became more *tryoni*-like it increased in its ability to tolerate high temperatures. Thus hybridization between species can produce the genetic variation necessary for natural selection to produce a new adaptive peak.

In many cases the potential evolutionary importance of inter-specific hybrids is eliminated because they are sterile. This is due to the failure of the two complements of chromosomes to pair properly during meiosis. However, several reproductive strategies allow both plants and animals to overcome this sterility and take advantage of the 'fixed heterozygosity' of hybrids. Several of these methods are associated with polyploidy (three or more sets of homologous chromosomes).

Polyploidy in plants

It has been estimated that at least one-third of angiosperms, and a higher proportion of ferns, are polyploid. In many cases these polyploids are thought to have arisen primarily by inter-specific hybridization. On rare occasions, spontaneous doubling of the F_1 hybrid chromosomes takes place, forming a fertile allopolyploid (amphiploid) with a tetraploid chromosome composition. As there are now two pairs of each parental chromosome type, meiotic pairing can take place. These allopolyploids are reproductively isolated from the original parents due to the production of sterile triploids in backcrosses, and thus form new species at a single step. Polyploidy is much more important in the origin of plant species than in animals because, by self-fertilization or vegetative reproduction, a single plant may give rise to a population which represents a new species. Self-fertilization and asexual reproduction, although occurring in some invertebrate phyla, are uncommon in vertebrates. Polyploids can be formed also by multiplication of the same chromosome set (autopolyploid). It is difficult to distinguish strictly between allopolyploidy and autopolyploidy, as the two genomes in any individual of an outbreeding sexually-reproducing species will be genetically distinct.

Studies using fraction 1 *protein*

Many agriculturally important plants are polyploid and, because of the potential value to breed improvement schemes, considerable effort has been devoted to determining the ancestral diploid parental species. Of particular importance in this respect have been the studies on fraction 1 protein (ribulose 1,5- diphosphate carboxylase-oxygenase). This protein is found in all organisms which have chlorophyll A, and it is the major soluble protein in the leaves of most plants. Fraction 1 protein is composed of eight high-molecular-weight polypeptides and eight low-molecular-weight polypeptides. It is very suitable for the determination of the origin of plant species which have arisen by interspecific hybridization for two reasons: (1) the protein is monomorphic in all species so far examined; (2) the large subunits are coded by chloroplast DNA and hence inherited solely from the maternal parent, since the gametes derived from the pollen are in the form of nuclei, unaccompanied by chloroplasts. On the other hand, the small subunits are coded by nuclear DNA and are inherited in normal Mendelian fashion. On isoelectric focusing in 8 M urea, the eight large subunits resolve into three general protein bands and the small subunits into one to four bands depending on the species. The three large subunits are possibly due to post-translational modifications, but the small sub-units are coded for by separate loci.

The genus *Nicotiana* comprises some 64 species, among which is the commercial tobacco plant *N. tabacum* which has a chromosome number ($2n$) of 48. This species is believed to have arisen by allopolyploidy after hybridization of *N. sylvestris* ($2n = 24$) with another species, variously suggested as being *N. tomentosa*, *N. tomentosiformis* or *N. otophora*.

Nicotiana digluta, a synthetic species known to have arisen by chromosome doubling following the hybridization of *N. glutinosa* and *N. tabacum*, was examined by Kung *et al.* (1975). The isoelectric focusing pattern of fraction 1 protein from *N. digluta* was found to be identical to that of the infertile F_1 hybrid, and showed three bands for the large subunit and four for the small one. The large subunits were of identical isoelectric points to those of *N. glutinosa*, the maternal parent in the original hybridization. The small subunit bands were a composite of those of *N. glutinosa* and *N. tabacum*, both of which had two fractions of different isoelectric points, thus indicating an equal contribution of maternal and paternal nuclear genes (figure 7.3).

A similar comparison of the fraction 1 polypeptide patterns from the putative progenitors of *N. tabacum* has indicated conclusively that *N. sylvestris* contributed the large subunit gene and therefore was the maternal parent in the original hybridization (Gray *et al.*, 1974). *N. sylvestris* also contributed one of the two small subunits, the other being introduced by *N. tomentosiformis*, the male parent (figure 7.4).

The oats (*Avena*) and the wheats (*Triticum*) consist of diploid, tetraploid and hexaploid species. In *Avena* there are two main groups of diploids which are assigned the genomic symbols A and C. Polyacrylamide electrophoretic comparisons have shown that the fraction 1 protein of the A genome diploids has a greater anodal mobility than the C genome diploids (Steer, 1975). This difference in mobility results from differences in the large subunits (Steer and Kernoghan, 1977). The single

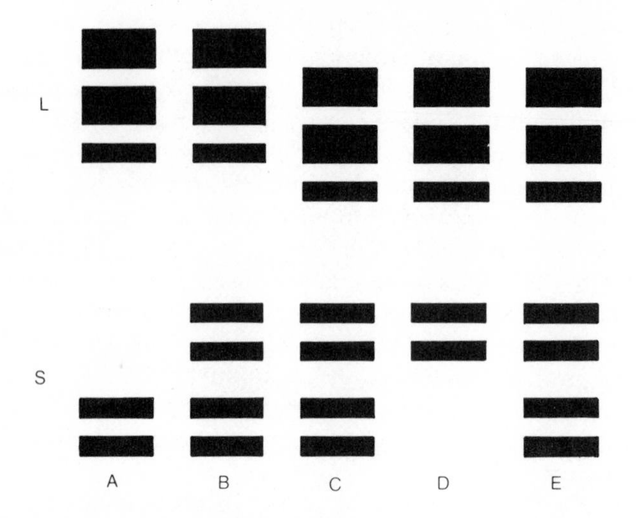

Figure 7.3. Polypeptide composition of fraction 1 protein from various *Nicotiana* species, as revealed by isoelectric focusing in gels containing 8 M urea.
A, *N. tabacum*; B, *N. tabacum*(♀) × *N. glutinosa*(♂); C, *N. glutinosa*(♀) × *N. tabacum*(♂); D, *N. glutinosa*; E, *N. digluta*. L = large subunit polypeptides; S = small subunit polypeptides (drawn from photograph in Kung *et al.*, 1975).

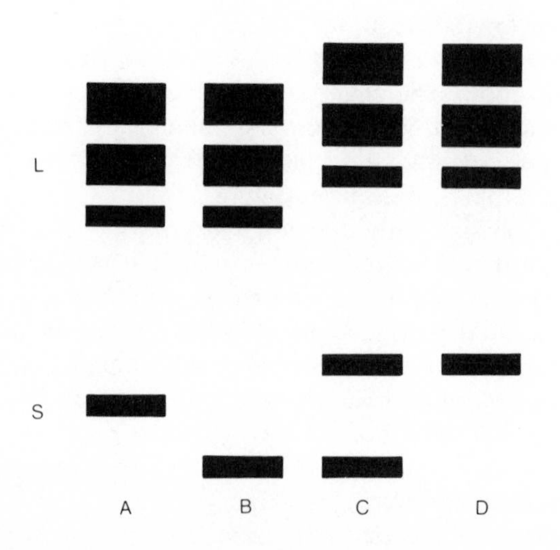

Figure 7.4. Polypeptide composition of fraction 1 protein of *N. tabacum* and the putative progenitor species, as revealed by isoelectric focusing in gels containing 8 M urea. A, *N. otophora*; B, *N. tomentosiformis*; C, *N. tabacum*; D, *N. sylvestris*. L = large subunit polypeptides; S = small subunit polypeptides (drawn from photograph in Gray *et al.*, 1974).

small subunit focuses at the same pI-value in all *Avena* species. This homogeneity of the small subunit means that fraction 1 protein can contribute information only on the maternal parent in *Avena* polyploids. From isoelectric focusing studies of the large subunit, it appears that the C genome diploids have not been involved (as the maternal parents at least) in the formation of either the tetraploid or hexaploid species of *Avena*.

Similarly, in the *Triticum* species, only a single type of small subunit is found on isoelectric focusing. Two distinct types of large subunit are present, and thus information can be obtained on the origins of the cytoplasmic genes (Chen *et al.*, 1975).

Enzyme patterns in polyploid plants

Electrophoretic studies have shown that tetraploid plants generally express enzymes of both diploid parents. In cases where the diploid parents had different alleles at a particular locus, the tetraploid exhibits a 'fixed heterozygote' pattern for that locus. The increase in enzyme forms in the tetraploid may be instrumental in extending the range of environments in which the species can live and may account for the wider distribution of many tetraploid species relative to their diploid progenitors. Polyploidy is much more common in colder environments. Thirty-eight percent of flowering plants are polyploid in the northern Sahara as compared with 71% in Greenland (Chai, 1976).

Gottlieb (1976) has reported on the enzyme variation in three diploid *Tragopogon* species and their naturally occurring tetraploid derivatives. The three diploid species, *T. dubius*, *T. porrifolius* and *T. pratensis*, were introduced to America during the present century and so it is known that the tetraploids are of very recent origin. Two tetraploid species have originated following allopolyploidy of F_1 hybrids. *Tragopogon miscellus* arose from the hybrid of *T. dubius* × *T. pratensis*, and *T. mirus* from *T. dubius* × *T. porrifolius*. Electrophoretic examination showed that the three diploid species are distinct in their allelic composition at about half of their loci, and consequently the tetraploid hybrids were substantially heterozygous, e.g. for the dimeric enzyme alcohol dehydrogenase (ADH), for which there are three loci in the diploid plants and thus six ADH isozymes, *T. miscellus* showed 13 isozymes out of the 15 theoretically possible.

Hybrid and polyploid animals

A number of clonally reproducing unisexual species of fishes, amphibians and reptiles have been discovered in recent years. Electrophoresis has been used to confirm that many of these are of hybrid origin.

Hybridogenesis in fish and amphibians

The all-female species of Mexican fishes of the genus *Poeciliopsis* have arisen as hybrids between sexual species of this genus. Each unisexual form contains a maternal genome derived from *P. monacha* and a paternal genome from one of the three species, *P. lucida*, *P. occidentalis*, or *P. latidens*, with which the unisexuals live and mate (Vrijenhoek *et al.*, 1977, 1978). The F_1 hybrid genotype of the unisexuals is maintained through the process of hybridogenesis in which the *monacha* chromosomes pass to the haploid egg during meiosis, but the paternal genome is discarded. Mating with the 'sexual host' species then restores the hybrid genotype of the zygotes. The maternal *monacha* genome passes unaltered from mother to daughter in a clonal fashion, whereas the paternal genome is obtained at random from the other species. The absence of recombination means that there will be variability in the paternal genes but not in the maternal ones, and the term *hemiclone* has been coined to describe this situation. Electrophoretic analysis has shown that there are at least eight hemiclones in the *P. monacha–lucida* 'species'. It seems likely that in most cases this is due to multiple hybrid origin, i.e. each F_1 zygote derived from mating between two sexual species establishes a distinct hemiclone. However, the possession of allozymes of two loci in a *P. monacha–occidentalis* population which were not present in sexually reproducing populations of *P. monacha* suggest that this hemiclone was produced by mutation in another hemiclone. These unisexual fish were found to have high levels of heterozygosity (.36–.50) in keeping with their inter-specific hybrid origin.

Summation of parental electrophoretic patterns has confirmed that the European water frog *Rana esculenta* is a hybrid between two distinct species, *R. ridibunda* and *R. lessonae* (Uzzell and Berger, 1975). *Rana esculenta* is hybridogenetic with two different breeding groups. One type discards the *R. lessonae* chromosomes at meiosis, and so it lives and mates with this species. The other type is similarly associated with *R. ridibunda.* Enzyme patterns have allowed the recognition of several hemiclones within each of the two types.

Gynogenesis

A further method of maintaining the hybrid genome is by gynogenesis, which is practised by a number of fishes, amphibians and reptiles,

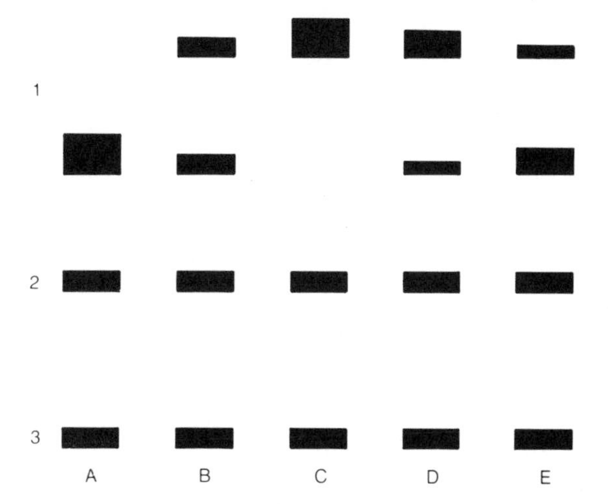

Figure 7.5. Muscle general protein patterns of *Poeciliopsis*. A, *P. lucida*; B, *P. monacha-lucida*; C, *P. monacha*; D, *P.* 2*monacha-lucida* (triploid with 2 *monacha* and 1 *lucida* genomes); E, *P. monacha*-2*lucida*. The depth of the fractions in group 1 is proportional to their staining intensity (drawn from photograph in Vrijenhoek, 1975*b*).

including triploid *Poeciliopsis* and diploid and triploid *Poecilia*. In gynogenesis 'fertilization' of the egg is necessary by a male of a bisexual species. The sperm does not, however, contribute any genetic material but serves merely to activate development of the egg.

In triploid *Poeciliopsis*, all three genomes are expressed and this is reflected in the relative quantities of different proteins in the electrophoretic patterns (figure 7.5). *Poecilia formosa* is an all-female species that arose by hybridization between *P. latipenna* and *P. mexicana*. The parental species differ in the electrophoretic mobility of their serum albumins and the hybrid *P. formosa* exhibits both types (Abramoff *et al.*, 1968). In triploids, one of the albumins is of greater intensity (in similar fashion to figure 7.5) and this has been used to investigate their occurrence in natural populations. Balsano *et al.* (1972) found that the ratio of diploid to triploid *P. formosa* individuals ranged from 1:0.24 to 1:34 in five populations sampled in north-eastern Mexico.

Parthenogenesis

A third method by which hybrids can reproduce and thus form separate species, is parthenogenesis (thelytoky). This is development without fertilization and is of two types:

1. *Apomixis*, in which divisions of the oocyte are by mitosis.
2. *Automixis*, in which meiosis is retained.

The parthenogenetic lizard species *Cnemidophorus tesselatus* is composed of diploid populations which are produced through the hybridization of the bisexual species *C. tigris* and *C. septemvittatus*, and triploid populations derived from the crossing of diploid *C. tesselatus* and a third bisexual species *C. sexlineatus*. Parker and Selander (1976) have analysed allozymic variation at 21 loci in samples of *C. tesselatus* from southern USA. They found considerably higher heterozygosity in *C. tesselatus* (0.560 in diploids and 0.714 in triploids) than in the parental bisexual species (mean 0.059). All triploid individuals apparently represented a single clone, but 12 separate clones were noted in the diploid hybrid. Mutation was suggested as the basis for the origin of three clones, multiple hybrid origin for four clones, and recombination for five clones.

Clearly then apomictic organisms, as a consequence of the absence of recombination, show very high levels of genetic variation, the heterozygosity of the original hybrid persisting in each generation. This high heterozygosity, coupled with the increased rate of reproduction due to parthenogenesis, probably accounts for their success.

Parthenogenetic species of a number of other animal groups are known, especially among the insects. In many of these cases parthenogenesis is associated with autopolyploidy. The parthenogenetic weevils have been studied in detail by Suomalainen, Saura and others. In a large number of cases, the weevil group consists of diploid bisexual species, each of which has a derivative polyploid parthenogenetic form. In Europe, the bisexual forms of these weevils are usually quite restricted in distribution, while the parthenogenetic ones are much more widespread, e.g. in the genus *Otiorrhynchus*, bisexual forms are found in limited areas of the Alps while their parthenogenetic equivalents are widespread in northern Europe. It has been suggested that the bisexual forms survived in glacial refugia, while the parthenogenetic ones were produced in, and colonized, the wider area after the retreat of the last glaciation.

Based on an electrophoretic study of enzymes in bisexual, and triploid and tetraploid parthenogenetic populations of *Otiorrhynchus scaber*, Suomalainen and Saura (1973) refuted the hybrid origin of these parthenogenetic forms. They also concluded that parthenogenesis had arisen as a single event, and that the differences among populations could be explained by mutations which occurred after the acquisition of parthenogenesis, since a majority of alleles present in the polyploid parthenogenetic races of *O. scaber* were also found in a single diploid bisexual population. The degree of heterozygosity in triploid and tetraploid populations was not significantly different from that in the bisexual population studied. This is as would be expected from an autopolyploid origin.

Bisexual polyploids

A number of bisexually reproducing polyploid animals are known and others have been suggested without definitive evidence. Ohno (1970) has proposed that polyploidy played a significant part in vertebrate evolution in the stage before the sex chromosomes became differentiated. Once this latter process was completed in the reptiles, birds and mammals, polyploidy seems to have been disruptive to the mechanism of sex determination. In most fish and amphibians, the sex chromosomes have not yet been established, and bisexual polyploids are known in a number of groups. In frogs, a number of diploid-tetraploid sibling species have been recognized (Bogart and Tandy, 1976). Teleost fish of the family Salmonidae and some Cyprinidae are regarded, on good evidence, as tetraploids. Salmonid fishes show duplication of many loci, e.g. they have at least five lactate dehydrogenase loci as compared with three in most other teleosts. Polyploidy brings about a duplication of all the genes, and in most cases is followed by functional diploidization, with the formation of divalent rather than tetravalent sets of chromosomes at meiosis.

Duplicated genes can have one of three fates:

1. The redundant copies may be 'silenced' by mutation to non-functional alleles (null alleles) or by regulatory mutations. This may be followed by physical loss of the redundant DNA.
2. Duplicated genes may diverge in structure and function, and be preferentially expressed in different tissues or at different developmental stages.
3. The same function may be retained, but with different alleles fixed at the two loci, resulting in permanent heterozygosity. This can provide heterosis without the genetic load which accompanies a single-locus polymorphism.

In a number of species of cyprinid and salmonid fish which are sufficiently old for more than 95% of duplicate loci to have diverged to the extent of producing electrophoretically distinct proteins, only some 35-65% of the genes are expressed in duplicate (Ferris and Whitt, 1977*a*, 1977*b*, 1977*c*: Allendorf *et al*., 1975). Allozyme studies have suggested that the other 65-35% of duplicate genes have been silenced rather than never having diverged.

From studies of a number of catostomid fish, Ferris and Whitt (1977*a*) have shown that those species which are morphologically divergent from the ancestral form have lost the expression of more duplicate genes than those with a more primitive morphology, as judged by comparison with fossil forms. An average of 59% of the loci showed evidence of duplicate gene expression in the morphologically conservative group, compared with 42% in the faster-evolving species.

The loss of duplicate gene expression is a potentially powerful method for elucidating the phylogeny of polyploid groups. A prerequisite for cladistic analysis is the identification of primitive and derived character states. Since by polyploidy all loci would have originally been duplicated, the ancestral condition is clearly defined. Each locus is a potential character and can exist in two character states—the ancestral or duplicate condition, and the derived or single-locus condition, as a result of the silencing of one locus. Ferris and Whitt (1978) found that phenetic and cladistic relationships derived from the levels of gene duplication in thirty species of catostomid fishes were in close agreement with those expected from traditional phylogenies based on morphological data.

Hybridization and the origin of domestic animals

No such elegant approach to the determinination of the hybrid origin of animals has been carried out as yet along the same lines as for tobacco fraction 1 protein. The demonstration, by restriction endonuclease analysis, that the mitochondrial DNA of horse × donkey hybrids (mule, hinny) is maternally inherited (Hutchison *et al.*, 1974), suggests that a similar approach is possible.

Information on protein polymorphisms in domestic cattle and sheep can be most adequately explained by assuming hybrid origins. Nearly all allelic variants of proteins differ by single amino-acid substitutions, e.g. of 145 human haemoglobin variants, only one (Hb C Harlem) differs from normal haemoglobin by two amino acids, and this involves a second substitution at a different site in the relatively common haemoglobin S. In the few cases where several amino-acid differences are found, these are the result of deletions or insertions rather than substitutions.

Manwell and Baker (1976) have drawn together the sequence data for polymorphic proteins in cattle and sheep. They found that, in about a third of protein variants in these animals, multiple amino-acid differences exist. In these cases, the variants differ by two, three, and in one case, by seven amino acids. Bridging intermediates were not found between any of these variants (Hb S is the bridging intermediate between Hb A and Hb C Harlem). These results suggest that domestic cattle and sheep may have evolved through hybridization of species or near-species.

If two species which differ by three amino acids in a particular protein hybridize, then the hybrid will be a heterozygote producing both these proteins. If the hybrid then forms a new species (artificially or naturally) these two variants will exist in that species, giving a polymorphism for that particular protein. Since, in general, different species differ by several amino acids in their orthologous proteins (chapter 8), this results in a polymorphism in which the variants differ by more than a single substitution.

Domestic sheep *Ovis aries* are polymorphic for haemoglobin, with two main types, A and B, which differ by seven amino-acid substitutions in

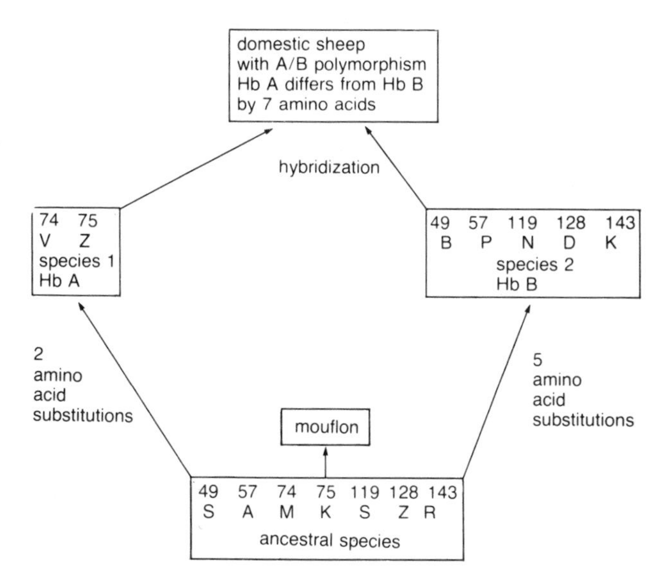

Figure 7.6. Evolutionary divergence for sheep haemoglobin β chain variants A and B, followed by hybridization to produce present-day domestic sheep. The mouflon β sequence is assumed to be the same as the common ancestor. Numbers refer to amino-acid sites in the β chain. Single letter amino-acid codes are given in table 2.1 (after Manwell and Baker, 1976).

the β chain. This is greater than the difference between sheep A and goat A. The amino-acid sequence of the β chain of the mouflon *Ovis musimon* is intermediate in its sequence between domestic sheep β^A and β^B. Manwell and Baker suggest that the mouflon β chain may have changed relatively little from the ancestral β chain. They postulate the divergence of three types from this ancestor: the mouflon and two sheep species which differentiated in their haemoglobins forming the A and B types respectively. These two then hybridized to form the present-day domestic sheep with the AB haemoglobin polymorphism (figure 7.6). With its higher oxygen affinity, the sheep A haemoglobin is better adapted to living at high altitude. Conversely, sheep with haemoglobin B are more adapted to lowland conditions. Thus it is postulated that these two forms evolved as adaptations to different environments. Summer-winter mountain migrations of man could have been responsible for bringing these two different sheep species together in the early days of domestication. Interestingly, wild Iranian sheep appear to have only the haemoglobin B variant, and so may represent a stock close to one of the ancestral species.

The apparently higher level of polymorphisms in domestic species may also be indicative of their hybrid origins. However, for economic and availability reasons, domestic species have been subjected to more intense research than wild ones, and this may have resulted in more polymorphisms being uncovered. A proper detailed comparison is necessary to establish if this is the explanation.

CHAPTER EIGHT

HIGHER-CATEGORY SYSTEMATICS

Amino-acid sequence studies

At the level of families and above, intra-specific variation is insignificant 'noise' and so one individual is an adequate representation of each taxon. This enables slow, but precise, methods such as amino-acid sequencing to be used. In most cases, for practical reasons, the purified protein is isolated from a pooled sample of a number of individuals. The development of methods for the determination of the primary structure of proteins has provided biologists with a powerful tool for investigating higher-category relationships and for constructing phylogenies.

Although conceptually this approach is simple, there are difficult practical problems involved, some of which have not yet been adequately resolved. First, comparisons must be made between orthologous proteins (p. 25). While, in many cases, when dealing with electrophoretic patterns it is clear which bands are orthologous and which paralogous, when an isolated protein is involved, the distinction may be more difficult to make. A comparison of the alpha haemoglobin polypeptide of one mammal species with the beta polypeptide of another would be meaningless in the determination of the relationship of the two species—the differences found would be those which had evolved since the original α–β duplication took place in the early fishes.

If amino-acid sequences of two orthologous polypeptides are aligned, then the number of amino-acid differences between them can be determined by comparison (position by position) along the length of the chains. A direct measure of genetic distance is thus obtained which represents the number of amino-acid replacements which have taken place since the taxa, from which the proteins were obtained, shared a common ancestor. Multiple pairwise comparisons can be used to determine evolutionary divergence within a group of organisms and from this information, phenetic and cladistic relationships can be deduced. Since orthologous proteins are present in such diverse organisms as bacteria, begonias, bees, birds and biologists, the possibility exists of determining the evolutionary history of all living organisms.

Evolutionary distances

Various methods have been developed to quantify the differences between orthologous sequences. A simple count of the number of amino acids is an underestimate, since two or more point mutations may be required for the interchange of certain amino acids. Using the genetic code, amino-acid sequences can be translated into messenger RNA or DNA codon sequences. Due to the redundancy of the code, this translation cannot be unambiguous since only methionine and tryptophan are represented by single codons. In practice, the *minimum mutation distance* is used, i.e. the minimum number of nucleotide substitutions required to convert the codon for one amino acid into the codon for the other. Thus a minimum of a single base change is required to convert the codon for aspartic acid into that for glutamic acid, but two changes are needed to alter it to a cysteine codon and three to the methionine codon (table 2.1).

Neither amino acid nor minimum mutation differences take account of multiple superimposed changes and of chance or convergent replacements. Thus a particular site may, in the course of time, be occupied by a number of different amino acids. Previous substitutions cannot be discovered from contemporary sequences. Various probability methods have been developed in an attempt to correct for superimposed changes. Dayhoff (1972) has calculated that for two sequences to differ by 50 amino acids in 100 sites, about 83 amino-acid substitutions must actually have taken place, 33 of which were superimposed. Based on this, Dayhoff has adopted a unit of change called the PAM (Accepted Point Mutation), which is the number of accepted point mutations per 100 amino acids. Thus two proteins differing by 50% of their amino acids have a genetic distance of 83 PAM units (table 8.1). The discrepancy due to superimposed changes is relatively small, and insignificant up to a sequence difference of about 20%.

In other cases the unit of evolutionary distance used is the corrected number of minimum base changes. Most methods for correcting for multiple substitutions give distances which correlate well with each other, and it is not clear at the moment which of these methods is the best one. Beyer *et al.* (1974) and Holmquist (1976) have discussed the desirable properties of distance measures.

Table 8.1 The relationship between the observed number of amino-acid differences per 100 amino acids of two sequences and the number that must have occurred, i.e. the evolutionary distance, is given. This represents the extent of superimposed and back mutations (based on Dayhoff, 1976).

Observed differences in 100 amino acids	*Amino-acid evolutionary distance in PAMs**
1	1.0
5	5.0
10	10.5
20	23.2
30	38.8
40	58
50	83
60	117
70	170
80	260
90	595

* Accepted point mutations per 100 amino acids

Sequence alignment

Due to deletions and insertions, orthologous proteins from different organisms may not show the same number of amino-acid positions. In this case alignment is not a straightforward operation but requires the introduction of gaps in one or other sequence in order to maximize the number of identical amino acids. This creates a problem in that if sufficient gaps are introduced in one or both of the sequences being compared, then some degree of correspondence will inevitably be detected even between non-homologous proteins. Prosthetic groups, e.g. haem, can be of considerable value in determining alignments.

The human α haemoglobin polypeptide consists of 141 amino acids and the β polypeptide of 146. Maximum similarity is achieved if they are aligned over 148 positions by the introduction of seven space positions in the α chain and two in the β chain (figure 8.1).

Systematic uses of protein sequences

The *Atlas of Protein Sequence and Structure* and its supplements (Dayhoff, 1972; 1973; 1976) contain some 500 sequences of 20 amino acids or more from a wide range of organisms, and many new ones are published annually in the biochemical journals. Undoubtedly, due to technical errors and incomplete sequencing, there are errors in a minority of the published sequences. Romero-Herrera *et al.* (1978) have noted that some earlier unexpected similarities in myoglobin sequences were in fact due to sequence errors. More weight can obviously be given

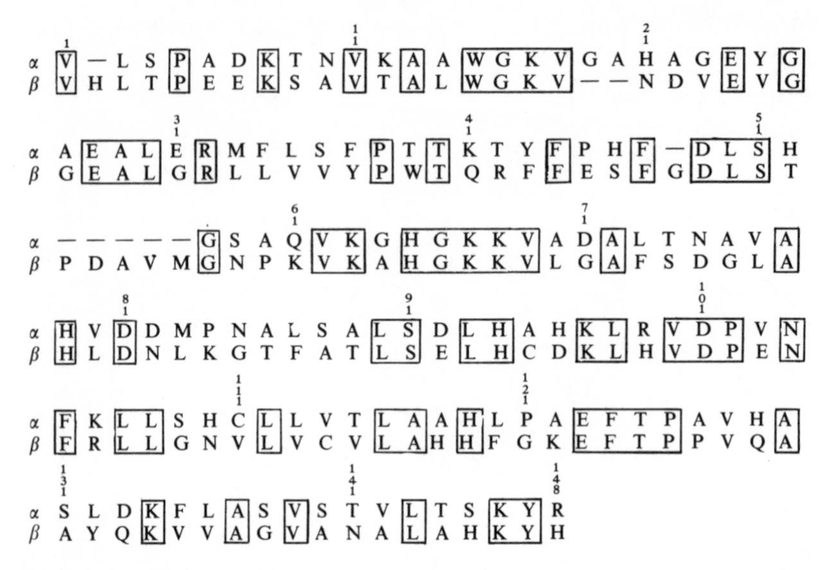

Figure 8.1. Alignment of α and β haemoglobin polypeptides of man over 148 positions. Single-letter amino-acid codes are given in table 2.1 (data from Dayhoff, 1976).

to a conclusion if it is supported by the sequences of several different proteins. The published sequences represent an enormous bank of information which can be applied to a number of systematic problems from genus to phylum level. However, the sequences so far available represent only a minuscule fraction of those that will be necessary before open questions such as the phylogenies of plants and invertebrates can be tackled. Of the presently available sequences, about half are from mammals and many of the remainder involve other vertebrates or bacteria. From the insect group, with probably more than 1 million species, only 8 sequences are contained in the *Atlas.*

Amino-acid or base differences among orthologous proteins or genes provide good estimates of phenetic relationships and can be analysed by the methods of numerical taxonomy. They can also be used to infer phylogenetic information, and considerable attention has been given to the construction of phylogenetic trees from such data. These trees indicate the branching order of the taxa and the amount of amino-acid or nucleotide change which has taken place in each branch. Phylogenetic trees produced from sequence data of a single protein type should not be considered as definitive statements of phylogenetic relationships. They form, however, hypotheses which are useful starting points for the consideration of a number of systematic and evolutionary questions.

Two main types of methods have been used for tree construction:

1. Matrix methods
2. Ancestral sequence methods

Phylogenetic trees can be constructed by the matrix methods already outlined for electrophoretic data, or by algorithms developed specifically to deal with protein sequence information. Many of these methods involve a 'trial and error' procedure to determine the tree which represents best the original information or which involves overall the least number of mutational events (i.e. most parsimonious). The maximum-parsimony method obeys the phylogenetic reconstruction principles of Hennig in that ancestral and derived character states are distinguished (Moore, 1976).

The ancestral sequence method of Dayhoff (1972) is based on amino-acid data. A computer algorithm is used and this generates the most probable ancestral sequence at each node (branching point) on the premise that a common ancestor is likely to have, in any given sequence position, that amino acid which is present in the majority of its closest known descendant and antecedent relations. After the initial tree is constructed, a 'branch-swapping' procedure is carried out to determine the most parsimonious tree.

At the moment, it is not clear which of the tree construction procedures is most suitable. All result in phylogenetic hypotheses which can then be tested in the light of existing and subsequent data.

Cytochrome c

The electron carrier cytochrome c is ubiquitous in the mitochondria of aerobic organisms. Its small size (103–112 amino acids), coupled with the relative ease with which it can be purified and its stability, facilitated the determination of its sequence in a wide range of organisms. So far, sequences for some 50 organisms have been published, and these range from protozoa to man. The sequence of cytochrome c_2 of the bacterium *Rhodospirillum rubrum* has also been determined, and it appears to be closely related to eukaryote cytochrome. Homology is evident when all the sequences are aligned, with organisms as remote as man and baker's yeast having identical amino acids in 64 positions. The chance of two such similar sequences arising by convergent evolution is very remote, and it is more likely that this similarity is due to evolution from a common ancestral gene. All cytochrome c sequences so far determined have the same amino acid at 26 sites (20 sites, if the *Rhodospirillum* c_2 is included). These are presumably the amino acids which cannot change without the function of the protein being damaged.

Cytochrome c has evolved at a comparatively slow rate and thus closely related organisms have the same sequence, e.g. man and chimpanzee; zebra and donkey; cow, sheep and pig; camel and whale; turkey and chicken; cauliflower and rape. Table 8.2 is a matrix showing

Table 8.2 Differences in the amino-acid sequences of the cytochromes c of various organisms. The numbers of differences between sequences are given above the diagonal and the percentage differences below the diagonal (based on Dayhoff, 1976).

	1	2	3	4	5	6	7	8	9	10	11	12	13	14	15	16	17	18	19	20	21	22	23	24	25	
1. Human and chimpanzee		10	10	13	15	14	18	18	20	29	27	44	43	44	46	46	46	42	44	43	42	43	40	41	48	1
2. Pig, cow and sheep	10		6	9	9	20	11	11	14	24	22	44	41	46	48	49	48	45	46	45	45	46	42	42	46	2
3. Grey kangaroo	10	6		12	11	21	13	13	17	26	24	45	44	46	49	49	48	44	46	47	44	45	42	45	47	3
4. Chicken and turkey	13	9	12		8	19	11	15	18	23	23	44	42	47	49	49	49	45	47	46	46	46	40	43	49	4
5. Snapping turtle	14	9	11	8		22	10	13	19	26	24	47	44	45	47	48	47	44	45	46	44	45	40	42	47	5
6. Rattlesnake	13	19	20	18	21		24	26	27	28	29	47	43	45	47	47	45	42	44	47	43	43	42	45	52	6
7. Bullfrog	17	11	13	11	10	23		13	21	28	22	46	44	48	50	51	50	46	48	48	47	47	43	45	48	7
8. Carp	17	11	13	14	13	25	13		12	26	21	44	41	47	49	50	51	45	47	47	46	46	45	45	45	8
9. Pacific lamprey	19	13	16	17	18	26	20	12		26	28	48	47	52	52	52	52	47	49	51	49	51	49	49	50	9
10. Garden snail	28	23	25	22	25	27	27	25	25		27	48	45	51	53	53	53	51	52	51	52	50	49	50	53	10
11. Screw-worm fly	25	20	22	21	22	27	20	20	26	25		44	36	46	46	49	49	43	44	45	44	42	42	45	54	11
12. Baker's yeast	40	40	41	40	43	43	42	41	44	44	41		36	47	51	49	51	47	47	46	47	47	46	48	61	12
13. *Neurospora crassa*	40	38	41	39	41	40	41	38	44	42	34	33		48	49	53	54	49	49	50	49	49	50	51	51	13
14. Rape and cauliflower	39	41	41	42	40	40	43	42	46	46	41	42	43		6	7	11	11	9	10	14	13	13	18	59	14
15. Hemp	41	43	44	44	42	42	45	44	46	47	41	46	44	5		8	12	11	11	15	13	14	14	20	60	15
16. Elder	41	44	44	44	43	42	46	45	46	47	44	44	48	6	7		6	11	10	13	11	15	14	18	60	16
17. Box-elder	41	43	43	44	42	40	45	46	46	47	44	46	48	10	11	5		13	13	13	12	14	15	18	61	17
18. Leek	38	40	39	40	39	38	41	41	42	46	39	42	44	10	10	10	12		11	15	12	15	12	19	57	18
19. Nasturtium	39	41	41	42	40	39	43	42	44	46	40	42	44	8	10	9	12	10		15	11	13	16	21	58	19
20. Wheat	38	40	42	41	41	42	43	42	46	46	40	41	45	9	13	12	12	13	13		15	14	17	17	60	20
21. Sunflower	38	40	39	41	39	38	42	41	44	46	40	42	44	13	12	10	11	11	10	13		13	15	20	59	21
22. Parsnip	38	41	40	41	40	38	42	41	46	45	38	42	44	12	13	14	13	14	12	13	12		16	21	60	22
23. Buckwheat	36	38	38	36	36	38	38	41	44	44	38	41	45	12	13	13	13	11	14	15	14	14		14	61	23
24. *Ginko biloba*	36	37	40	38	37	40	40	40	43	44	40	42	45	16	18	16	16	17	19	15	18	19	12		61	24
25. *Euglena gracilis*	46	44	45	47	45	50	46	43	48	51	50	56	47	53	54	54	54	51	52	54	53	54	54	54		25

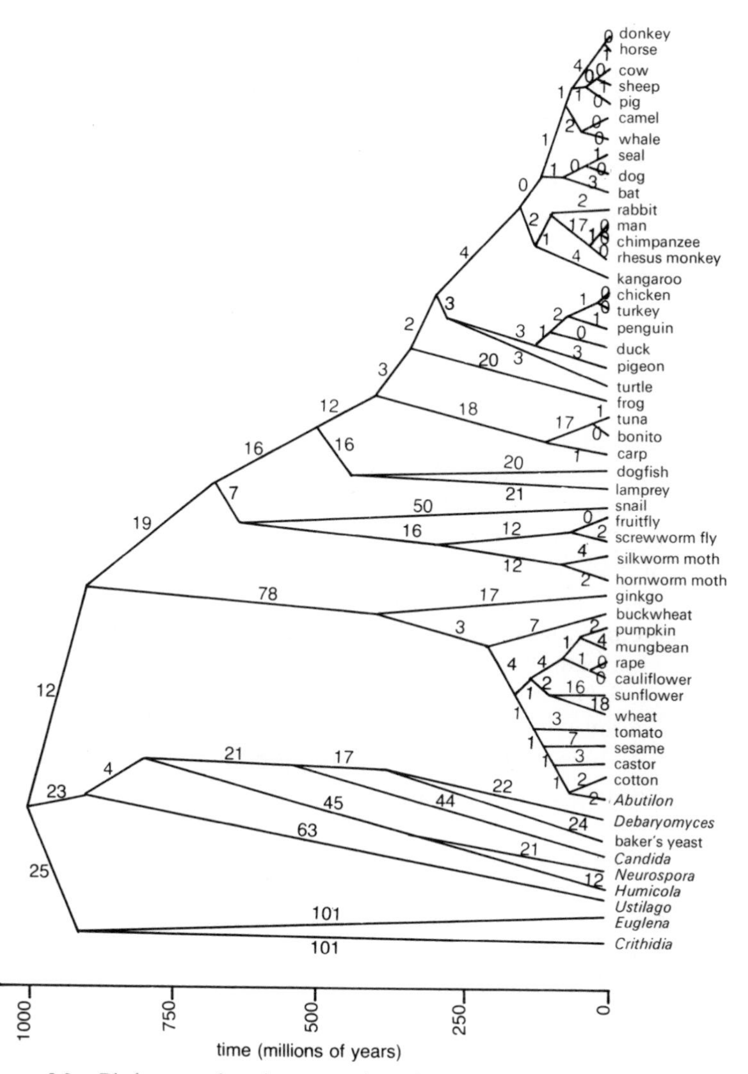

Figure 8.2. Phylogeny of various organisms based on amino-acid differences in 53 cytochromes c. The numbers on the branches are the corrected numbers of mucleotide substitutions estimated to have occurred between the various branching points (redrawn from Moore *et al.*, 1976).

the percentage differences in amino-acid sequences of cytochrome c between various pairs of organisms. Figure 8.2 presents some of these results in the form of a phylogenetic tree. It most cases, this information is in keeping with that expected from conventional classification and phylogeny, and it is remarkable that this should be so on the basis of a single protein. There are, however, some points of disagreement, e.g.

chicken is closer to penguin (2% sequence difference) than it is to pigeon or duck 3–4%). Similarly the turtle is closer to the birds (8%) than it is to another reptile, the rattlesnake (21%). The determination of whether these and other discrepancies are due to convergence or to errors in current phylogenies will require evidence from sequences derived from many more taxa and other proteins. Lyddiatt *et al.* (1978) have studied the cytochromes c sequences of certain annelids, molluscs and echinoderms, and found that the phylogeny derived from these is not in agreement with currently accepted views on invertebrate evolution.

Other proteins

Cytochrome c shows a low rate of amino-acid replacement compared with many other proteins. This slow rate of evolution makes it particularly suitable for examining relationships at the class and phylum level, but of little value for more closely related taxa. The next most sequenced proteins after cytochrome c are the vertebrate globins— α and β polypetides and myoglobin. Globins evolve about four times faster than cytochrome c, which make them suitable for application to lower-level problems. Figure 8.3 shows the phylogeny of the globin genes based on paralogous comparisons of the α and β polypeptides and myoglobin, and also of various vertebrates from orthologous comparisons. Again the globin tree is, in most respects, in close agreement with a phylogeny based on morphological and other evidence. Important also is the concurrent independent evidence given by the α and β polypeptides. The rat *Rattus norvegicus* shows a difference of 21 amino acids in the α chain from that of the mouse *Mus musculus*, thus questioning the close relationship of these two species and their placement in the same subfamily, Murinae.

For studies below the family level, faster-evolving proteins are available, e.g. the fibrinopeptides which show an amino-acid replacement rate some twenty times greater than cytochrome c. The soluble protein fibrinogen is a constituent of vertebrate plasma. In blood clotting, fibrinogen is converted, by partial proteolysis, into insoluble fibrin. This process involves the release of fibrinopeptides A and B. Figure 8.4 is a phylogenetic tree of the artiodactyl order of mammals based on sequences of fibrinopeptides. Once again this is in good agreement with conventional placing.

Amino-acid sequence analyses may seem to be of little systematic value if they merely confirm relationships known from other studies. It is important, however, to test the utility of amino-acid sequence analyses for higher-category systematics in groups such as the vertebrates, which have a relatively good fossil record, before this approach is extended to

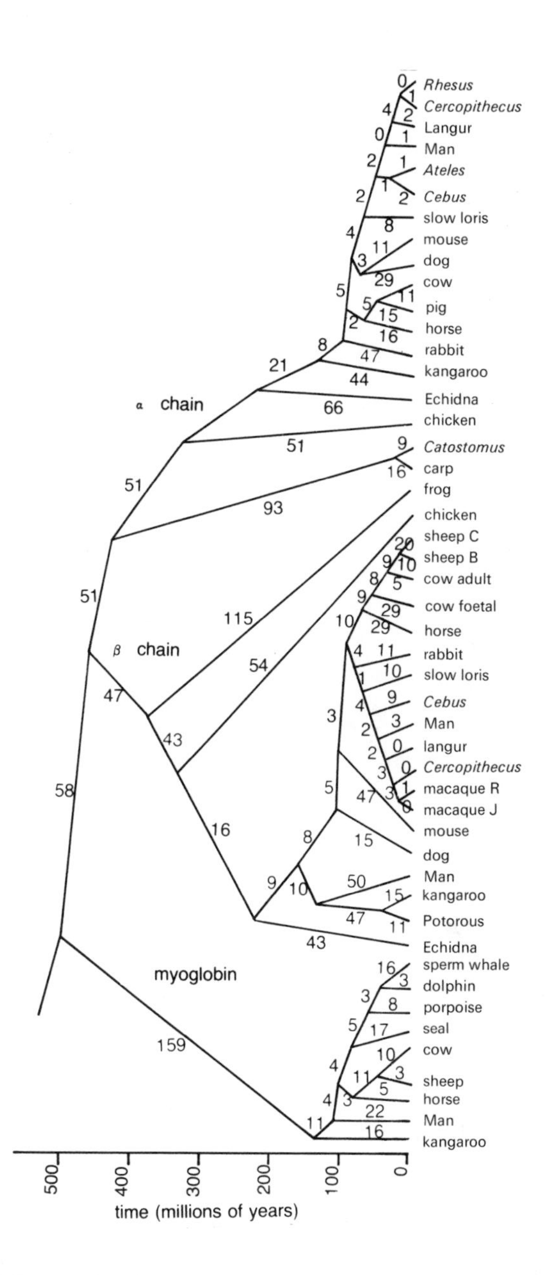

Figure 8.3. Phylogeny of various organisms and of three globin genes based on amino acid differences in 48 globins. The numbers on the branches are the corrected numbers of nucleotide substitutions estimated to have occurred between the various branching points (redrawn from Goodman, 1976).

invertebrates and plants where the fossil record is generally poor or absent. Amino-acid sequencing is a slow and expensive procedure as well as requiring relatively large amounts of purified protein. Now that automatic sequencing equipment is available, this latter aspect is probably the greater hurdle, e.g. to obtain 5 mg of purified cytochrome c the processing of from one to ten kg of starting tissue may be required. A glance at figures 8.2 and 8.3 will show that many of the taxa so far examined are those which are readily available to the biochemist, i.e. laboratory animals

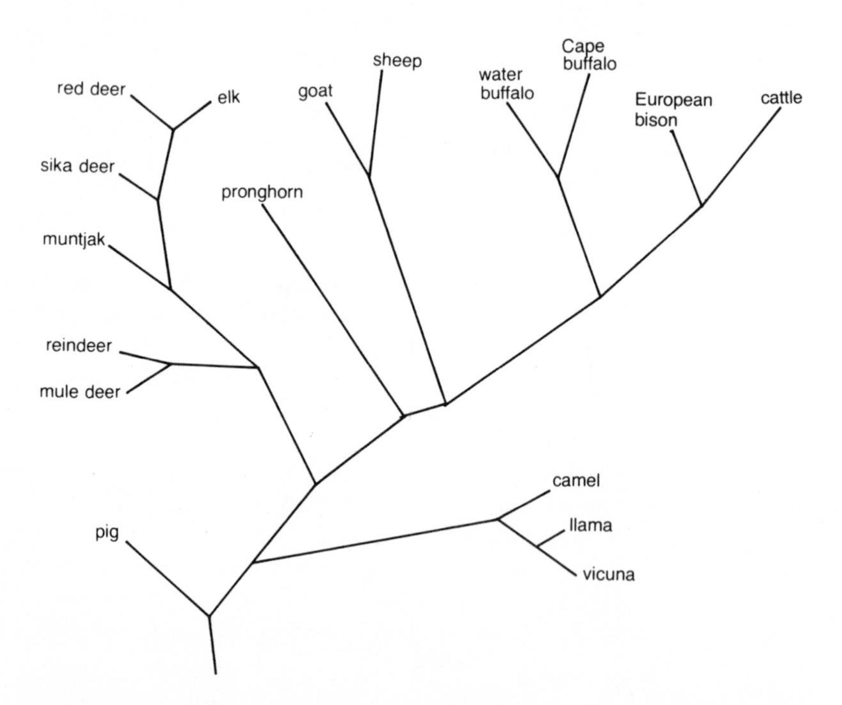

Figure 8.4. Phylogenetic tree of some artiodactyls based on amino-acid sequence comparisons of fibrinopeptides. It is almost identical with the currently accepted tree for this order, based on morphological data (redrawn from Mross and Doolittle, 1967).

and those obtainable from biological supply-houses or local markets. Only in a few cases have sequences been determined in an attempt to solve systematic problems. For significant systematic information to become available, many more sequences will be necessary. Of particular importance in this respect are the recent series of plant cytochrome c and plastocyanin sequences produced by Boulter's group at Durham University (see Haslett *et al.*, 1978, and references cited therein). Fortunately there are several short-cut methods which can be used in higher-category work.

Micro-complement fixation

The micro-complement fixation technique (p. 30) provides an indirect measurement of the number of amino-acid differences between orthologous proteins. Using proteins of known amino-acid sequences, it has been shown that the approximate relationship between immunological distance y and percentage sequence difference x is $y = 5x$ for albumin, transferrin and bird lysozyme, while for pancreatic ribonuclease it is $y = 7x$. Using plant plastocyanins of known sequence, Wallace and Boulter (1976) found a similar relationship.

This technique, then, provides a speedy and reliable indirect way of determining sequence differences. It does not, of course, indicate the specific amino-acid replacements, only the overall number. Immunological distance can be used in a similar fashion to genetic distance derived from electrophoretic studies for the examination of systematic problems, the production of phylogenies, and the estimation of divergence times. Sarich (1977) compared Nei's genetic distance with immunological distance for 76 pairs of species and found a strong correlation between these two measures ($r = 0.82$).

Wilson and his associates have used the micro-complement fixation technique to estimate sequence differences in albumin and other proteins in a wide range of vertebrate species. Albumin is very suitable protein for such systematic studies, as it evolves relatively rapidly and consists of a single polypeptide of about 580 amino acids. It is thus equivalent in size to the combination of cytochrome c, myoglobin, α and β haemoglobin polypeptides, and A and B fibrinopeptides. In one study, these workers examined the clawed frogs of the genus *Xenopus*. Since these frogs are widely used in biological research, a correct taxonomy is particularly important. They found, however, that frogs commonly used by North American molecular and developmental biologists under the name of *Xenopus muelleri* were in fact another species, *X. borealis* (Bisbee *et al.*, 1977).

Maxson (1977) has used the steady rate of albumin evolution to detect a probable case of convergent evolution involving the frog *Anotheca spinosa*. On morphological grounds, this monotypic genus is assigned to the subfamily Amphignathodontinae along with the marsupial tree frogs. However, based on albumin immunological distance, *Anotheca* was found to be as similar to several North American *Hyla* species as these were to each other. Also *Anotheca* albumin was very different from the albumins of the marsupial tree frogs—as different, in fact, as these were from *Hyla* (figure 8.5). The albumin information would indicate that the true taxonomic position of *Anotheca* is as a member of the subfamily Hylinae. The placement of *Anotheca* in the Amphignathodontinae is

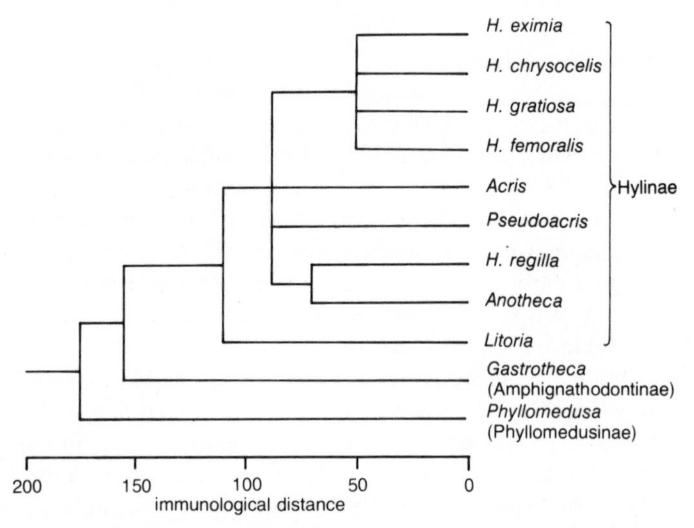

Figure 8.5. Phylogenetic relationships among various tree frogs of three sub-families as deduced from micro-complement fixation studies of albumins. Note the position of *Anotheca* as a member of the hyline assemblage (redrawn from Maxson, 1977).

probably erroneous due to convergent evolution by which it morphologically resembles the marsupial tree frogs.

A similar case of convergent evolution involves the frogs *Hyla wrightorum* and *H. regilla.* These two species are close enough in their external morphology to have been classified as the same species by some workers. On immunological grounds they are as different, on average, as *Hyla* and the related genus *Acris*, and appear to be members of two distinct lineages.

According to most ornithologists, the pheasant is more similar in morphology and way of life to the chicken that it is to the turkey. This view is reflected in bird classification, where traditionally the pheasant is put together with the chicken in the subfamily Phasianinae of the family Phasianidae, whereas the turkey is assigned to another family, the Meleagrididae. However, micro-complement fixation studies of nine proteins (Nolan *et al.*, 1975) showed pheasant-chicken differences to be the same or greater than turkey-chicken differences.

The phylogenetic position of the giant panda *Ailuropoda melanoleuca* has been the subject of a long-standing controversy among taxonomists. The main point at issue is whether it is closer to the procyonids (raccoon, etc.) than to the bears, albeit both groups are fairly closely related. Using antiserum to both albumin and transferrin, Sarich (1973) has confirmed the giant panda is a highly specialized bear. The transferrin immunological distance between the giant panda and various bears was of a similar magnitude to that found between other quite closely-related species pairs, e.g. cat and lion, dog and fox.

Using antiserum against 1-glycerophospate dehydrogenase, Collier and MacIntyre (1977) determined the immunological distances among 34 species of drosophilids. They found a good correlation between immunological distance and the classical determinations of these species. They also found that immunological distances of the order of four to six units could be detected between allozymes from the same species, which probably reflects a single amino-acid substitution.

Wallace and Boulter (1976) compared immunological distances among 37 species of higher plants by means of antisera to the plastocyanins of spinach *Spinacia oleracea* and nettle *Urtica dioica*. While there was a general similarity between immunological distance and classical taxonomic distance, they noted some exceptions, e.g. some species which were members of different families showed greater immunological similarities than some members of the same family. It is not known whether these differences reflect changes in evolutionary rates in different lineages or errors in the existing classification.

DNA hybridization

Since its inception in 1960, the technique of DNA hybridization (p. 46) has been used widely to investigate systematic relationships among viruses, bacteria, plants, and animals. Much of the early work on plants and animals examined hybridization of repeated DNA sequences. However, it has been shown that divergence can occur, between members of families, of repeated DNA sequences, even within a single species (Shields and Straus, 1975). Consequently, interspecific comparisons could be obscured by intra-specific divergences accumulated over a much longer period of time. Sequences which are present only once per haploid genome are much more reliable for systematic work. The techniques involved are time-consuming and expensive, due in part to the need to label the DNA from one species with a radioactive isotope. It appears to be quite difficult to obtain consistent results with this technique, and large differences can be found in reciprocal comparisons.

Shields and Straus (1975) found in their study of various birds, that variation in melting temperatures of DNA from six species of *Junco* was inside the range found in the homologous reaction. Thus within the experimental error, there were no significant differences in the unique DNA of this particular genus. More distantly related species did exhibit lowered thermal stabilities (figure 8.6).

Results of comparisons of several groups of species by DNA hybridization are summarized in table 8.3. It is clear from this table that the percentage of nucleotide differences (as estimated by the reduced thermal stability of the hybrid) is lower for closely related species. Thus

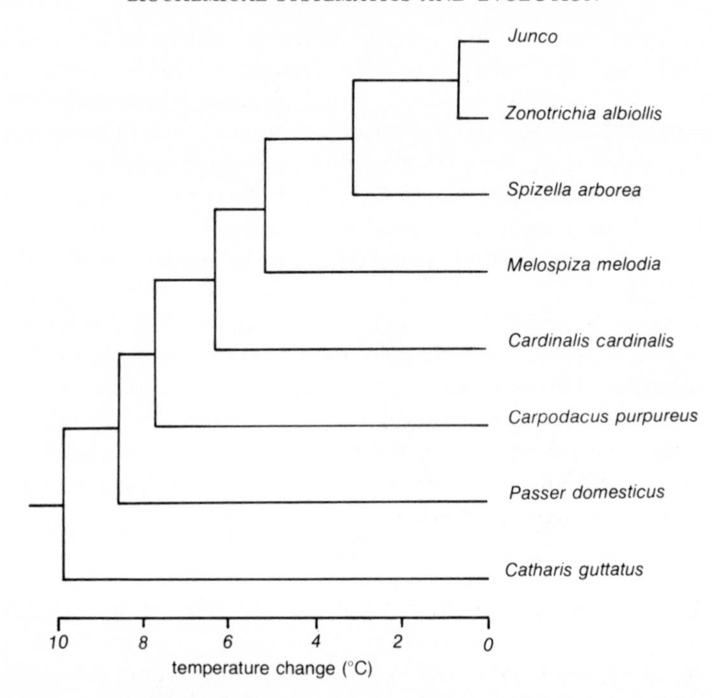

Figure 8.6. Phylogenetic relationships of several species of birds based on differences of melting temperatures of homologous and heterologous unique sequence DNA-DNA hybrids. The temperature difference is approximately equal to the percentage of nucleotide substitutions between the various taxa (after Shields and Straus, 1975; and Dobzhansky *et al.*, 1977).

Table 8.3 Comparisons of genetic similarities of various organisms by the technique of DNA hybridization (data for mammal comparisons (except *Mus*) from Kohne *et al.* (1972), others from Wilson (1975)).

Taxa compared	*Percentage differences in sequences of non-repetitive DNA**
man–chimpanzee	1.6
man–gibbon	3.5
man–rhesus monkey	5.5
man–galago	28.0
mouse–rat	20.0
cow–pig	20.0
cow–sheep	7.5
Mus musculus–M. cervicolor	5.0
Xenopus laevis–X. borealis	12.0
Simulium pictipes–S. venustum	11.0
Drosophila melanogaster–D. salmon	19.0

* Assuming a 1% difference in base sequence lowers thermal stability by 1°C.

the proportion of nucleotide pairs is 1.6% between man and chimpanzee, while it is 28% between man and galago. In general, nucleotide differences estimated from DNA hybridization are much higher (about × 5) than those inferred from amino-acid sequences. This could be due to several factors, e.g. mutations at the third codon position do not usually affect protein structure, and the rate of these may be higher than at other positions. As already mentioned, technical difficulties with the DNA hybridization technique may account, in part, for the observed discrepancies.

Time and molecular evolution

The availability of information of the nucleic-acid and amino-acid differences among species, from application of the above techniques, has led biologists to look for a relationship between the extent of the sequence difference and the time of divergence of various species pairs as originally determined from the fossil record. Surprisingly, the initial studies appeared to suggest that a given protein evolves at nearly constant rate even in diverse lineages. This opened up the possibility that if the rate of evolution for a particular protein could be determined from fossil or other evidence in one group, it could be used as a 'molecular clock' in other groups where the fossil evidence is absent or equivocal. Since different proteins evolve at different rates, an appropriate 'clock' could be chosen for the study of distant or of recent cladogenic events. Before examining the reliability and use of individual molecular clocks, a consideration of the variability among proteins in their rate of evolution is appropriate.

Evolutionary rates can be quantified in a number of ways. The most commonly used measures are:

1. Accepted point mutations per 100 amino acids (PAM units) (p. 132).
2. The rate of substitutions per amino-acid site per year, expressed in Pauling units where 1 Pauling = 10^{-9}.
3. Unit evolutionary periods (UEP) which is the time needed for a 1% difference in amino-acid sequences to show up between divergent lines.

Evolutionary rates for some proteins are listed in table 8.4. It has been suggested that the rate of protein evolution depends both on the probability that a substitution will be compatible with the biochemical function of the protein and on the dispensability of the protein to the organism. Although every amino-acid site in a protein may be presumed to have a function, the sites probably differ from one another in respect of the number of alternative amino acids that can carry out the function at each site. In the case of the active sites of enzymes or hormones, possibly only a single amino acid is suitable. By contrast, at other sites in a protein, two or more amino acids may be equally suitable. The high evolutionary rate of the fibrinopeptides has been suggested to be due to the fact that these carry out structural roles, holding the soluble fibrinogen molecule in a particular conformation. Any one of a number of amino-acid combinations may form a polypeptide suitable for that purpose. A good example of this functional constraint hypothesis is provided by the proinsulin molecule (Wilson *et al.*, 1977). In this molecule the C polypeptide acts as a bridge between the A and B polypeptides of proinsulin. Its function is to ensure proper pairing of the A and B chains during the formation of the insulin molecule. Once the latter has been formed, the C peptide is removed enzymatically. Experiments have shown that a variety of synthetic bridges can take the place of the C polypeptide in bovine insulin. Again it might be expected that various amino-acid combinations might be equally suitable. Consistent with this lax requirement, it has been found that the C polypeptide has evolved about seven times faster than the hormonal part of the insulin molecule (table 8.4).

The dispensability of a protein may also be important in governing its rate of evolution, i.e. whether or not there is an alternative pathway by which the same reaction can be carried out, or even if it can simply be dispensed with altogether, e.g. lysozyme appears to function in defence against bacterial invaders. Vertebrates, however, have many other ways of coping with invading bacteria.

Table 8.4 Rates of protein evolution. The unit evolutionary period (UEP) is the average time (in millions of years) needed for a 1% change in amino-acid sequence to show up between two divergent lines. The higher the UEP value, the slower the evolutionary rate (based on Wilson *et al.*, 1977).

Protein	*Unit evolutionary period*
Histone (H4)	400
Glutamate dehydrogenase	55
Collagen (α-1)	36
Lactate dehydrogenase (B^4)	19
Lactate dehydrogenase (A^4)	13
Cytochrome c	15
Insulin (hormone)	14
Insulin (C peptide)	1.9
Plastocyanin	7
Myoglobin	6
Haemoglobin (α polypeptide)	3.7
Haemoglobin (β polypeptide)	3.3
Albumin	3
Lysozyme	2.5
Fibrinopeptide A	1.7
Fibrinopeptide B	1.1
Immunoglobulin γ chains (V region)	0.7

After the neutralist–selectionist controversy, the constancy of orthologous protein evolution is probably the most hotly debated topic of molecular evolution. The two subjects are of course related in that, if most amino-acid substitutions are selectively neutral, then a time-dependent rate of fixation is easily explained. Thus the rate of evolution will depend only on the rate of occurrence of neutral mutations. Indeed, one of the strongest pieces of evidence for the neutralist theory is the constancy of protein evolution in very different organisms. If amino-acid replacements are subject to selection, then it is difficult to envisage how rates could be the same in species with very different internal and external environments, and with different population sizes.

Molecular clocks

For events to be used as an evolutionary clock, they must occur with regularity and be measurable accurately. Calibration of the clock must be possible from some external time measurement, e.g. a divergence of two species at a time which is known from the fossil record or from geological information. If this time is known, and the difference in macromolecular structure determined by amino-acid sequencing, micro-complement fixation, or DNA hybridization, then the change per unit of time can be calculated.

The first requirement is that the change in nucleic acids or proteins is measurable with sufficient accuracy. Although amino-acid sequences may be stated exactly, they are subject to errors in determination. This is particularly so when part of a sequence is determined from the total amino-acid composition and comparison with an orthologous polypeptide. Fitch (1976) has suggested that the error in a completely sequenced protein may be of the order of 2% and at least twice that when 'short-cuts' have been used. Micro-complement fixation and DNA hybridization are both subject to experimental errors, the extent of which has been noted already.

Macromolecular change can be determined by the above techniques with reasonable accuracy. The other requirement is for constancy of change. Fitch (1976) examined seven polypeptides which had been sequenced in 17 mammals, and plotted the combined total of nucleotide substitutions against the time of divergence of various pairs of species. He found (figure 8.7) a good overall correlation between time and rate of evolution, although there were significant deviations from constancy, especially in the primates. Fitch calculated that the variance in evolutionary rate is twice that expected of a purely random or Poisson process such as radioactive decay.

In order to calibrate protein clocks, accurate times of species

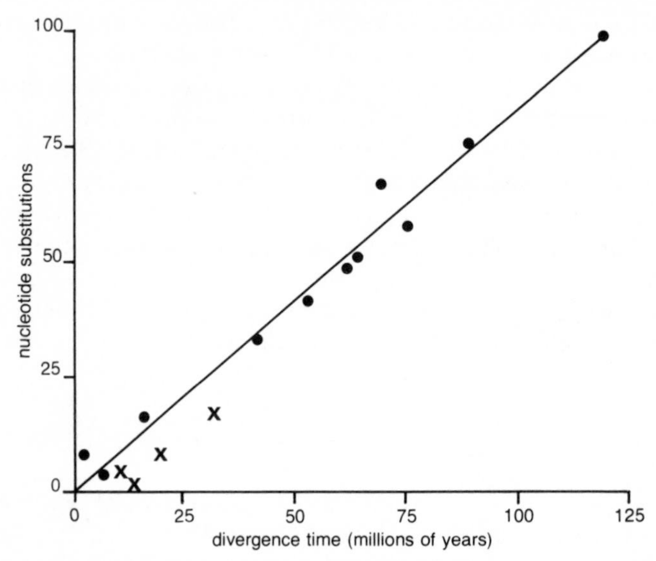

Figure 8.7. Nucleotide substitutions versus palaeontological time. The total nucleotide substitutions for seven polypeptides (cytochrome c, fibrinopeptides A and B, α and β haemoglobin, myoglobin and insulin C peptide—total of 578 amino acids) are plotted for pairs of mammals which separated from a common ancestor at the times indicated on the abscissa. The line was drawn through the outermost point, which represents the marsupial–placental divergence. The fit of points to this line is fairly good except for the primates (indicated as X) where protein evolution seems to have occurred at a slower rate than in other mammals. It should be noted that considerable uncertainty exists with regard to the divergence times of primates. Thus averaged over long periods of time and a number of proteins, nucleotide substitution would appear to be linear with respect to palaeontological time (redrawn from Fitch, 1976).

divergence are necessary. Most times are derived from fossil evidence. However, while palaeontologists can estimate the age of a fossil with accuracy, the assignment of it to a particular lineage is much more difficult. Wilson *et al.* (1977) have drawn attention to the need to differentiate between palaeontological fact and speculation.

Maxson *et al.* (1975) have used continental drift as an independent check on albumin evolution. The marsupial mammals and tree frogs of the subfamily Hylinae are found both in Australia and the New World. Plate tectonic studies have indicated that South America and Australia were connected by way of Antarctica until about 70 million years ago. Disappearance of this connection would have simultaneously isolated the marsupials and tree frogs in the two regions where they are now found. Regardless of the actual time of isolation, if albumin evolves at the same rate in both groups, then American marsupials should show a similar degree of immunological distance from Australian marsupials as do the respective groups of tree frogs. Maxson *et al.* found that this was indeed the case. Also the values obtained were in keeping with those expected from other studies for a separation of 70 million years.

A number of objections to, and apparent inconsistencies with, the constant-rate hypothesis have been raised. In the main, these may be countered by uncertainties in the palaeontological information (Wilson *et al*., 1977). One interesting example derives from attempts to relate molecular evidence to fossil evidence with regard to the origin of man. Palaeontologists have suggested 30 million years ago as the time of divergence of the human lineage from that of the African apes (i.e. chimpanzee and gorilla). From studies of sequence evolution in other mammals, species isolated for 30 million years would be expected to differ by about 5% in their amino-acid sequences and by about 8% in their unique DNA sequences. However, human proteins differ from those of the chimpanzee by about 0.8% of their sequences and for DNA there is some 1.6% difference. These values would indicate a divergence time of five million years. This discrepancy could be due to either a slowing in molecular evolution in the higher primates, or to erroneous dating of the time at which the two lineages diverged. However, as Wilson *et al*. (1977) have pointed out, the fossil evidence is consistent with a human–ape divergence of anywhere between four and 30 million years, with indisputable hominid evidence stretching back only 3.1 million years. By comparison of protein differences of non-mammal vertebrates with primates on the one hand and non-primate vertebrates on the other, they also demonstrated that while some proteins evolved faster in one lineage than the other, the overall rate of change was about the same. Thus molecular evolution does not appear to have slowed down in the primates.

In other cases, inconsistencies in evolutionary rate have been shown to be due to paralogous comparisons, e.g. bird lysozymes (p. 26). It seems that in a relatively few cases such anomalies represent changes in evolutionary rate in particular lineages. Goodman (1976) has suggested that the rate of amino-acid substitution in the globins has changed during the evolution of the vertebrates. He argues that earlier workers who concluded that globins evolved at a uniform rate had failed to account sufficiently for superimposed mutations. Goodman noted that the fastest rate of globin evolution occurred during the early evolution of the vertebrates subsequent to the myoglobin-haemoglobin duplication (p. 23). After the α–β duplication, which preceded the teleost–tetrapod separation, the α gene initially evolved more rapidly than the β gene, but between the tetrapod and amniote ancestors the rate of β evolution increased several fold. Between the amniote ancestor and the bird and mammal divergences, an abrupt slowing of both α and β evolution occurred. The rates again increased in the early mammals. Goodman accounts for these rate changes as follows.

The early increases in evolutionary rate were a result of the myoglobin–haemoglobin and α–β duplications. This latter duplication allowed the formation of a tetrameric molecule with a sigmoid oxygen equilibrium curve as a result of allosteric interactions. Following the development of a new form of a protein, the possibilities for improvement are probably at their maximum. Thus accelerated amino-acid substitution would have occurred, while selection was optimizing the structure of the new haemoglobin molecule, following both the initial duplication and the α–β duplication. Once the protein reached its optimum structure, the evolutionary rate decreased. Thus proteins which acquired their optimum structure at an early stage in evolution, e.g. cytochrome c, evolve at fairly constant rates in extant organisms. Wilson *et al.* (1977), however, have challenged Goodman's conclusions on the grounds that the divergence times used are speculative.

One example of changing evolutionary rates, which is independent of divergence times, concerns isozymes of aspartate aminotransferase. In animals, two isozymes occur, one in the cytosol (cAAT) and the other in mitochondria (mAAT). Using antisera to pig and to chicken, Sonderegger and Christen (1978) have determined the immunological distances among various vertebrates by means of micro-complement fixation. They found that the evolution of the two isozymes has proceeded at identical and constant rates throughout the evolution of the non-mammalian vertebrates. After the emergence of the mammals, however, the rate of the cAAT evolution more than doubled, while the mAAT continued to evolve at the same rate as in the non-mammalian vertebrates.

In conclusion, it appears that some proteins evolve at relatively constant rate over long periods of time, whereas in others, major changes in evolutionary rate probably take place. The average rates of protein evolution taken over long periods of time may be used as an approximate 'molecular clock'. However, the several sources of error preclude any naive use without detailed investigation of the protein concerned. As with other studies involving macromolecules, if several proteins give the same time estimate, this can be regarded with greater confidence. It is worth noting that macromolecular evolution appears to be dependent on absolute time and not on the number of generations per unit of time, as proposed by some workers. Also the requirements for the determination of phylogeny are considerably less stringent than the requirements for a 'molecular clock'. Protein differences would reflect phylogeny, even if changes in evolutionary rate have taken place, provided the rates at any given time are the same in different lineages.

CHAPTER NINE

COMPARISON OF BIOCHEMICAL AND MORPHOLOGICAL SYSTEMATICS

EARLIER CHAPTERS HAVE SHOWN THAT IN MANY CASES THERE IS a close similarity between classifications based on morphological and other traditional data and those derived from biochemical studies. In other situations divergent classifications are produced. Many morphologically identical or sibling species are quite distinct in their proteins, while some reproductively isolated species are biochemically similar. To what extent then do present techniques allow an examination of the molecular basis of phenotypic evolution? As a first approach to this question, consider the extent of genetic change which is necessary for speciation. A direct estimation of this aspect is complicated by the genetic change which takes place concurrently with, and independently of, the acquisition of reproductive isolation, and also by differences accumulated since speciation took place.

As tables 5.1 and 6.1 indicate, there is a general increase in genetic distance as taxa of increasingly higher rank are examined. Thus conspecific populations are genetically more similar than different species, and so on. There are, however, exceptions to this general trend.

Turner (1974) has studied allozyme variation at some 38 loci in five closely related species of North American pupfish (*Cyprinodon*). Although these species are morphologically, behaviourally and ecologically quite distinct, they are genetically as similar ($\bar{I} = 0.894$) as some conspecific populations.

Evolution of man and chimpanzee

Man *Homo sapiens* and chimpanzee *Pan troglodytes* show major differences in their structural features, ecology and behaviour, and are classified in two different primate families, the Hominidae and the Pongidae respectively. They are probably the best-studied pair of species at the macromolecular level. King and Wilson (1975) have compiled protein and nucleic-acid differences between humans and chimpanzees, based on studies of primary structure, micro-complement fixation, electrophoresis and DNA hybridization. All techniques yielded concordant estimates of genetic relationship. Since this is probably the only pair of species to be examined by all these techniques, this concordance is particularly interesting. Most of the sequenced proteins of man and chimpanzee are either identical or differ by one amino-acid substitution. On average, their proteins differ by about 0.8% of their sequences. Human–chimpanzee unique DNA hybrid dissociates at a temperature about 1.6°C lower than homologous DNA, and this is equivalent to about a 1.6% nucleotide sequence difference. Micro-complement fixation studies of serum albumin have given an immunological distance of six units. The genetic distance between man and chimpanzee, based on electrophoretic comparisons of proteins encoded by 44 loci, was calculated as 0.62. Although this is 25 to 60 times greater than the genetic distance between human races, it is relatively small and corresponds to the genetic distance between some sibling species of *Drosophila* and distinct congeneric species of other groups (figure 9.1). Comparisons with other primates suggest that about the same amount of DNA and protein change has taken place in the two lineages. There

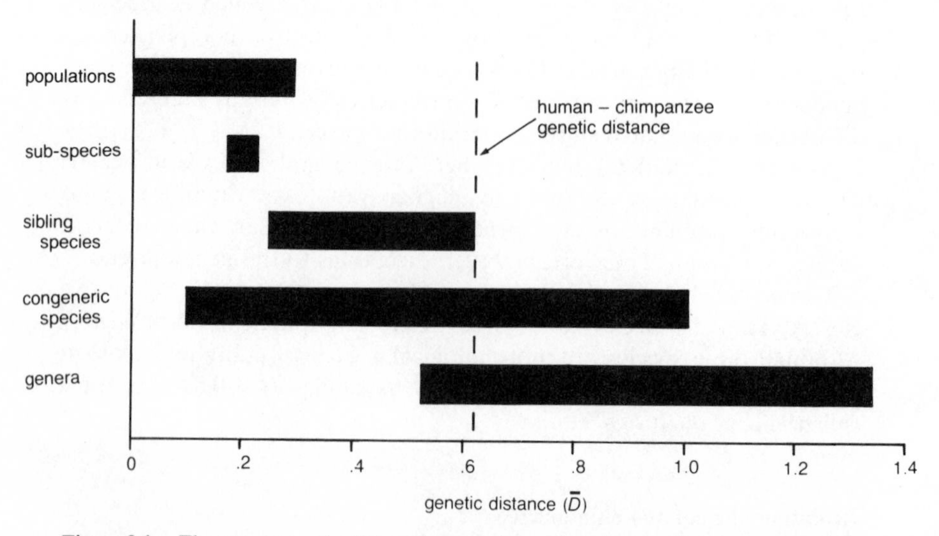

Figure 9.1. The mean genetic distance among taxa of various ranks, as determined by electrophoretic comparisons of proteins. The line indicates the genetic distance between humans and chimpanzees compared with that of other organisms (data from tables 5.1 and 6.2).

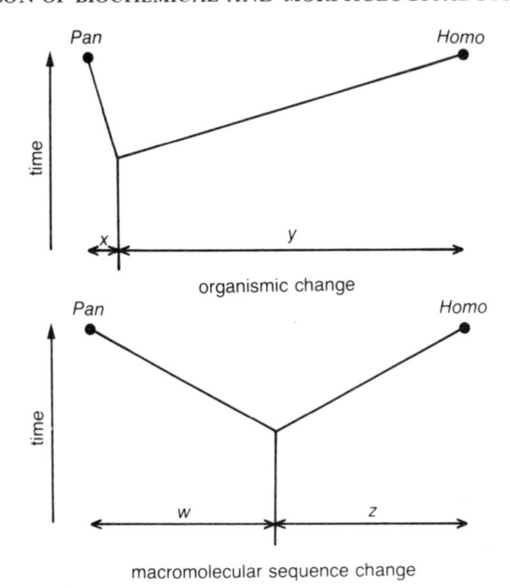

Figure 9.2. The contrast between organismal evolution and structural gene evolution since the divergence of the human and chimpanzee lineages. As the top diagram shows, more biological change has taken place in the human lineage *y* than in the chimpanzee lineage *x*. Protein and nucleic-acid evidence indicates that as much change has occurred in chimpanzee genes *w* as in human genes *z* (redrawn from King and Wilson, 1975).

appears, therefore, to have been a considerable difference between the rate of morphological evolution and that of macromolecular evolution in these two species (figure 9.2). However, as Grant (1977) points out, the differences between the two species are, in part, the result of cultural evolution. He suggests that 'the Hominidae could not be maintained separate from the Pongidae on morphological grounds alone, and, accordingly, we should reduce the two groups in taxonomic rank when we are discussing their strictly biological differences'.

Morphological versus genetic variation

Relatively few studies have involved detailed quantitative and objective comparisons of inter-specific differences in morphology with macromolecular differences. Partly this is due to the problems involved in quantifying some morphological differences. Schnell *et al.* (1978) have compared the use of meristic and enzyme characters in the classification of the kangaroo rats (*Dipodomys*). They found that essentially no correlation exists between morphological and genetic results when the taxonomic analysis is carried out by phenetic methods. Mickevich and Johnson (1976) have shown a similar lack of concordance between morphological and enzyme information in the fish genus *Menidia*, when evaluated by phenetic methods, but found close agreement when a cladistic approach was taken to the two character sets. The fact that both sets of information yield a similar phylogeny can be explained on the basis that when speciation takes place, divergence can then occur in both morphological and protein characters. Congruence in branching order does not require that the rate of morphological change following speciation be dependent upon the rate of protein change.

A study of covariation of genetic heterozygosity, as indicated by proteins, and morphological variation in the killifish *Fundulus heteroclitus*, has indicated that individuals heterozygous for an enzyme locus are likely to be less morphologically variable than individuals homozygous for that locus (Mitton, 1978).

There are a number of living organisms which, when compared with fossil relatives, appear to have remained virtually unchanged in their morphology for many millions of years. *Limulus polyphemus* closely resembles fossils more than 200 million years old. Analysis of some 24 loci in four populations of this species (Selander *et al.*, 1970) has shown that the proportion of polymorphic loci per population and heterozygosity are similar to values for other invertebrates. Similarly in the club-moss *Lycopodium lucidulum*, Levin and Crepet (1973) found a high level of variability, even though this species is morphologically similar to Devonian lycopods.

Morphological and macromolecular evidence can be seen to be discordant for three main reasons:

1. Either set of information may be incomplete or incorrect due to one or more of the possibilities outlined in earlier chapters.
2. Convergent evolution can result in similarity of particular characters in unrelated taxa.
3. Morphological and protein characters may evolve independently.

If, as discussed in earlier chapters, many amino-acid substitutions are selectively neutral, then such mutations cannot be the basis of adaptive evolution. The remarkable uniformity in the rates of amino-acid substitution in a particular protein has been noted also, and this again contrasts with the saltatory nature of much evolution at the morphological level. Biochemical and traditional classifications may only concur in those cases where morphological evolution has been strongly time-dependent. Thus structural gene evolution will appear to be correlated with organismic evolution if both are correlated with time.

Structural and regulatory genes

Since current studies on proteins examine only a small fraction (about 0.1%) of all structural genes, it is possible that those which are responsible for organismic evolution are to be found among structural genes which are not amenable to biochemical study. Alternatively, evolution of morphological features may be independent of structural gene changes and rely on modifications of regulatory gene systems. Consequently, altered patterns of gene expression may be responsible for adaptive evolution.

Small differences in the time of activation or in the level of activity of a single gene could in principle influence considerably the systems controlling embryonic development. Regulatory changes may be of at least two types. First, mutations could affect regulatory genes in a similar fashion to structural genes. Second, the order of genes on a chromosome may change owing to inversion, translocation, addition or deletion of genes, as well as by fission or fusion of chromosomes. Since the relative positions of genes are crucial, these gene rearrangements may have important effects on gene expression. Evolutionary changes in gene order are known to have occurred frequently. Microscopic studies of *Drosophila* chromosomes show that, in general, species do not have the same gene order and that inversions are the commonest type of gene rearrangement.

When a bacterial population encounters a new carbon compound, the primary event permitting the utilization of this new resource is a regulatory mutation which allows the production of a large quantity of an existing enzyme with a low activity towards the novel substance. Subsequent gene mutations may bring about a more efficient enzyme. Wilson *et al.* (1977) also point out that there are large concentration differences in certain proteins, even from closely related species (table 9.1). These concentration differences are possibly associated with major evolutionary adaptations.

Table 9.1 Differences in concentration of specific proteins. The concentration found in the taxon on the left of each pair exceeds that in the taxon on the right by an order of magnitude or more (based on Wilson *et al.*, 1977).

Protein	*Taxa compared*	*Tissue*
alcohol dehydrogenase	mouse/rat	kidney
amylase	man/dog	salivary gland
arginase	mammal/bird	liver
lactate dehydrogenase (A^4)	pheasant/petrel	breast muscle
	man/ruminant	liver
lysozyme c	man/cow	mammary gland
	man/rabbit	tear gland
	chicken/goose	oviduct
lysozyme g	goose/penguin	oviduct
myoglobin	whale/man	skeletal muscle
ribonuclease	ruminant/man	pancreas

Comparison of evolutionary rates in frogs and mammals

The vertebrates provide an ideal opportunity to compare the rates of evolution in genetic and structural features, as some lineages have evolved faster in morphology and behaviour than others, e.g. although there are thousands of species of frogs living today, they are so uniform phenotypically that they are placed in a single order, the Anura, even though they have been evolving along independent lines for almost 150 million years. Placental mammals on the other hand are divided into 16–20 orders. The morphological variation represented by mice, men, bats and whales, is without equal in the frogs, even though the latter are probably twice as old as the mammals. Based on micro-complement fixation measurements, albumin seems to have evolved at a constant and similar rate in both placental mammals and frogs (Maxson and Wilson, 1975; Maxson *et al.*, 1975). Thus species which are similar enough in anatomy and way of life to merit taxonomic distinction only at the species level (e.g. *Rana*) can differ as much in their albumins as does a bat from a whale, members of separate orders. Other proteins, such as haemoglobin, give similar results to that obtained from albumin (Baldwin and Riggs, 1974; Maxson and Wilson, 1975).

Inter-specific hybridization potential

Embryonic development involves an orderly programme of gene expression and requires that the two zygotic genomes contributed by the egg and sperm respectively be expressed simultaneously. Hybrid zygotes between distantly related species often show signs of breakdown in gene regulation as evidenced by allelic repression and asynchrony (p. 118).

It is of interest to consider whether the degree of protein similarity between parental species is correlated with the probability of successful development of inter-specific zygotes. The more similar the proteins of two species, the more likely it is, one might suppose, that their genomes would be compatible enough to permit development of viable hybrids. On the other hand, if compatibility of gene expression is essential for hybrid zygote development, then if regulatory evolution has proceeded slowly in frogs relative to mammals, it is reasonable to expect that frog species should retain the ability to hybridize with one another much longer than mammals. Because the rate of albumin evolution in frogs has been equivalent to that in mammals, small albumin differences should be found among mammals capable of hybridization and much greater differences between hybridizable frogs. Wilson *et al.* (1977) have shown, by micro-complement fixation measurements on albumin, that the average differences between pairs of mammals capable of producing viable hybrids was 3 immunological units in the 31 pairs examined, but 36 units for frogs (50 pairs examined) (figure 9.3). Using the known rates of albumin evolution, it can be estimated that it takes about two million years for a distance of three units to arise and about 21 million years for 36 units.

Thus mammals appear to have lost the potential for interspecific hybridization about ten times faster than have frogs. This is consistent with the suggestion that regulatory changes affecting embryonic

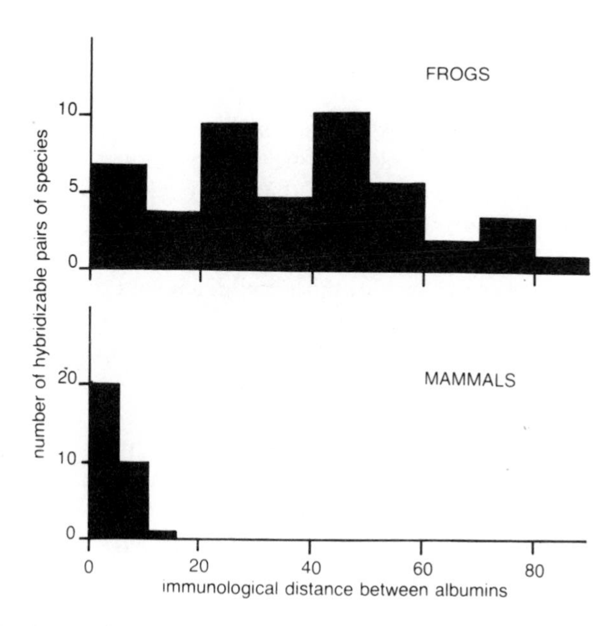

Figure 9.3. Immunological distances between albumins of species pairs capable of producing viable hybrids, as obtained by micro-complement fixation studies. Thirty-one such pairs of placental mammals and 50 pairs of frog species were investigated (redrawn from Wilson, 1976).

development have been acquired much more often in mammalian evolution than in frog evolution. Another possibility is that immunological reactions in placental mammals will lead to antibody formation and subsequent abortion of hybrid foetuses when the parents differ to a relatively small extent in their protein composition. Such reactions would not occur in frogs, which have external fertilization and embryonic development. Thus the hypothesis of rapid regulatory evolution in mammals is attractive, but much more evidence is required, and methods of studying regulatory gene evolution need to be developed.

Chromosomal studies

Since the relative position of genes on a chromosome is important in gene regulation, chromosomal mutations may bring about altered patterns of gene expression. Chromosomal changes such as fusion, fission, deletion and inversion result in changes in the number of chromosomes, and changes in individual chromosome morphology. These alterations may be observed when the chromosome complement of a cell is examined microscopically (karyotype analysis). Chromosomal changes can act as sterility barriers due to incorrect pairing in the heterozygote at meiosis. Such changes can result in reproductive isolation (and hence speciation) with little or no change at the structural gene level.

In the Israeli mole rat *Spalax ehrenbergi*, four chromosomal types ($2n = 52$, 54, 58, 60) are found, having largely allopatric distributions. Reproductive isolation exists among these four types, although in some areas of contact hybridization occurs between forms which differ by only two chromosomes (i.e. 52 with 54, and 58 with 60). Hybrids are rare and hybrid zones are narrow, the chromosomal differences providing effective postmating isolation. These four types of *S. ehrenbergi* may be considered as sibling species which have arisen by rapid speciation due to chromosomal mutations (stasipatric speciation).

Based on allelic variation at 17 loci, the mean genetic identity ($\bar{I}$) was found to be 0.978 (Nevo and Shaw, 1972). This is about the same degree of genetic relatedness as is observed between most conspecific populations (cf. table 5.1).

In the North American pocket gophers *Thomomys talpoides*, reproductive isolation has also been brought about by chromosomal rearrangements. The mean genetic identity ($\bar{I}$) for six chromosomal types was found to be 0.925, again of a level to be expected between conspecific populations. Rapid speciation as a result of chromosomal changes has also been described in plants of the genus *Clarkia* (Gottlieb, 1976). Reproductive isolation can likewise be achieved by autopolyploidy without any change in the structural genes. Clearly then, speciation can take place without any major changes occurring at the structural gene level and without long periods of allopatric isolation (see White, 1978, for detailed discussion).

Placental mammals seem to have experienced far more rapid karyotypic change than have frogs (Wilson *et al.*, 1974). The albumin immunological distance at which there is a 50% chance that two species will differ in chromosome number is about 6 units for mammals and about 120 units for frogs (figure 9.4). The rates at which chromosome arm number have changed were found to be very similar to the change in chromosome number (table 9.2). It is inferred that these two distinct

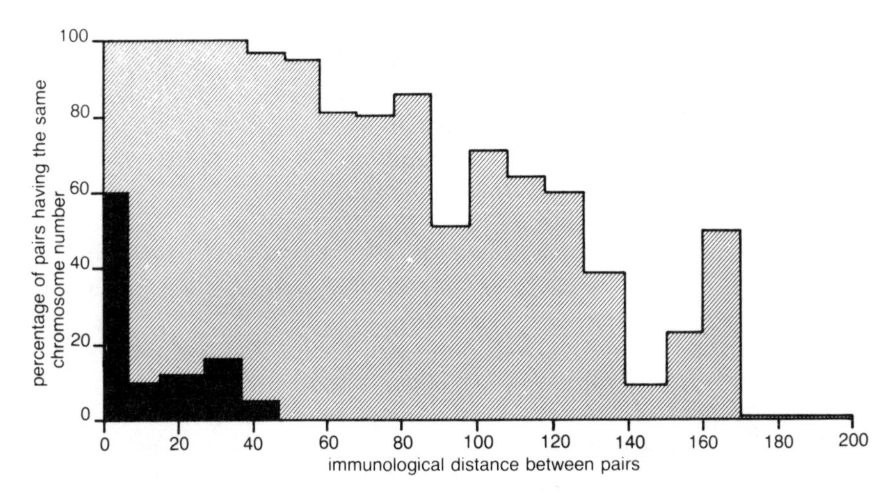

Figure 9.4. Proportion of species pairs having identical chromosome number as a function of the immunological distance between the albumins of the pairs. The light histogram summarizes the results for 373 different pairs of frog species. The darker histogram summarizes the results for 318 different pairs of mammal species (redrawn from Wilson, 1976).

Table 9.2 Rates of karyotype evolution in vertebrate genera (based on White, 1978).

	Number of genera examined	*Average age of genera (millions of years)*	*'Karyotypic changes' per lineage per million years*
horses	1	3.5	1.395
primates	13	3.8	.746
lagomorphs	3	5.0	.633
rodents	50	6.0	.431
artiodactyls	15	4.2	.561
insectivores	7	8.1	.187
marsupials	15	5.6	.176
carnivores	10	12.9	.078
bats	15	9.0	.059
whales	2	6.5	.025
Average for mammals:		6.5	.429
snakes	14	12.1	.048
lizards	16	20.1	.058
turtles and crocodiles	14	45.2	.022
frogs	15	26.4	.023
salamanders	11	23.4	.014
teleost fish	12	5.7	.029
Average for non-mammal vertebrates:		22.15	.032

types of gene rearrangement have each been evolving an order of magnitude faster in placental mammals than in frogs. There are, however, many types of chromosomal rearrangements which do not involve a change in chromosome number and it is possible that some of these may have occurred more frequently in frogs, compensating in a sense for the rarity of chromosomal fissions and fusions.

Wilson *et al.* (1975) have suggested that the high rates of karyotypic change in placental mammals are the result of social structuring in populations of these animals, which arises ultimately from a dependence of the young on the mother. This dependence, together with polygamy and dominance hierarchies, results in small populations and inbreeding which may promote rapid fixation of chromosomal changes and hence rapid speciation.

The above studies on the rates of evolution in frogs and mammals, summarized in table 9.3, suggest that changes in structural genes occur independently of organismic change. Regulatory changes, on the other hand, as manifested by studies of hybridization potential and karyotype, evolve in parallel with organismic changes. Studies on other groups also lend support to this hypothesis.

Table 9.3 Comparison of the rates of evolution in frogs and placental mammals (based on Wilson, 1975; and Wilson *et al.*, 1977).

Property	*Frogs*	*Placental mammals*	*Mammal rate/ frog rate*
Number of living species	3050	4600	
Number of orders	1	16–20	
Age of the group (millions of years)	150	75	
Rate of organismic evolution	Slow	Fast	3–20
Rate of albumin evolution	Standard	Standard	1
Rate of loss of hybridization potential	Slow	Fast	10
Rate of change in chromosome number	Slow	Fast	11
Rate of change in number of chromosome arms	Slow	Fast	14

Conclusions

In view of the independence of protein and phenotypic evolution, studies of proteins may seem to some to be of little systematic value. However, earlier chapters have reviewed many examples of the success of this approach. Where organismic evolution has occurred at a 'standard rate', protein evidence may correlate well with morphological evidence, because both have proceeded in a regular fashion with regard to elapsed time. This may be the situation where speciation has been allopatric, and in groups other than mammals.

The adoption of protein-based classifications in place of organismic classifications would lead to problems in the ranking of taxa. Protein differences are greater between geologically older species, such as frogs compared with the placental mammals. On protein criteria alone, the current scheme of placental mammal classification is greatly inflated relative to that of frogs. If a biochemical classification were to be adopted, basic changes in the ranking of either mammalian or anuran taxa would be required. Either the orders of placental mammals would have to be reduced to the rank of genera, or the genera of frogs be elevated to the rank of orders. The hope that proteins might ensure greater uniformity in the taxonomic ranking of organisms is unfulfilled.

The questions of whether and how the two types of information should be expressed in classifications are difficult to answer. They are related to the question, discussed in chapter one, as to whether classifications should be based on phenetic, cladistic or phylogenetic information. There is no *a priori* reason to suggest that classifications based on morphological features are more valuable than those derived from comparative protein and DNA studies. This re-opens the question of the functions of classifications. To some biologists, a classification based on phenotypic characters will be appropriate, whereas for the physiologists or molecular-orientated biologist, a biochemical classification may provide greater information. No single classification can hope to encompass these diverse sets of information. For many purposes, a quantitative statement of similarities and differences in morphology, behaviour or molecular structure may be more valuable than any shorthand classification which reflects only an intuitive arrangement of taxa. Molecular evidence has certain advantages and disadvantages, morphological evidence has others.

CHAPTER TEN

ANALYSIS OF ELECTROPHORETIC DATA

Calculation of allelic frequencies

The frequency of an allele is given by

$$\frac{2H_o + H_e}{2N}$$

where H_o = number of homozygotes for that allele
H_e = number of heteroygotes for that allele
N = number of individuals examined.

The polymorphism typified by figure 10.1 can be explained by assuming two-codominant alleles at the locus controlling the synthesis of this protein. It is conventional to label alleles in order of their decreasing electrophoretic mobility. Thus individuals of electrophoretic pattern type 1 are AA homozygotes; 2, AB heterozygotes; 3, BB homozygotes.

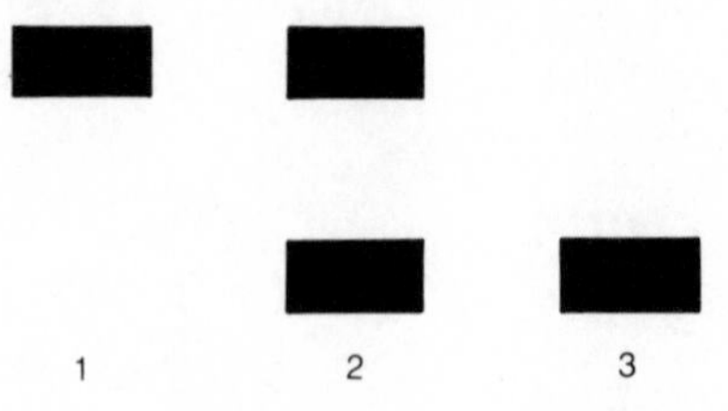

Figure 10.1. Electrophoretic phenotypes of three individuals (see text for genetic hypothesis).

Example 1

In the examination of 100 individuals, if the following numbers of these three phenotypes (≡ genotypes) are observed:

22 AA, 49 AB and 29 BB

then the frequency of allele A is

$$\frac{2 \times 22 + 49}{200} = 0.465$$

The frequency of allele B $= 1 - f_A = 0.535$.
The 'standard error' of the frequency of an allele (p) is estimated by

$$\sqrt{\frac{p(1-p)}{2N}}$$

Heterozygosity

The calculated heterozygosity per locus is

$$H_L = 1 - \Sigma x_i^2$$

where x_i is the frequency of the ith allele at a locus.

In example 1, $H_L = 1 - (0.465^2 + 0.535^2) = 0.497$.

The mean heterozygosity per locus $\bar{H}_L$ is the sum of H_L over all loci (including monomorphic loci where $H_L = 0$) divided by the total number of loci examined. The variance (V) of H_L for r loci is

$$\frac{\sum (H_m - \bar{H}_L)^2}{r(r-1)}$$

where H_m is the heterozygosity at the mth locus.

The standard error of $H_L = \sqrt{V}$.

The observed heterozygosity is the fraction of heterozygous individuals. In example 1

$$H_L(\text{obs}) = 49/100 = 0.49.$$

If the population is in Hardy–Weinberg equilibrium, then the calculated and observed heterozygosities will be very similar.

The calculated heterozygosity can be used also for asexual organisms, although in this case it bears no relationship to the observed number of heterozygotes. Nei (1975) has suggested that in this context it should be called *gene diversity* and not heterozygosity.

The mean heterozygosity per individual ($\bar{H}_I$) is the mean, over all individuals, of the proportion of loci at which each individual is heterozygous. Thus if an individual is heterozygous at 4 out of 10 loci examined, its heterozygosity (H_I) is 0.4. The values of $\bar{H}_L$ and $\bar{H}_I$ are the same, but their standard errors are different.

Effective number of alleles

The effective number of alleles is given by

$$\frac{1}{\Sigma x_i^2}$$

For example 1, the effective number of alleles is

$$\frac{1}{0.465^2 + 0.535^2} = 1.99$$

Hardy–Weinberg distribution

If a population is in Hardy–Weinberg equilibrium (p. 60), then the frequencies of genotypes will be in the ratio of p^2, $2pq$ and q^2, for a two-allele polymorphism, where p is the frequency of allele A and q is the frequency of allele B.

In example 1, the proportion of AA homozygotes should be ${f_A}^2 = 0.216$; of AB heterozygotes, $2 \times f_A \times f_B = 0.498$; of BB homozygotes, ${f_B}^2 = 0.286$. In 100 individuals the expected numbers of the three genotypes are given in table 10.1, along with the numbers actually observed.

Table 10.1 Observed distribution and expected Hardy-Weinberg equilibrium distribution of genotypes for example 1.

	Genotypes		
	AA	AB	BB
observed	22	49	29
expected	21.6	49.8	28.6

The difference between the observed and expected values can be tested for statistical significance, using a χ^2 test for goodness of fit. Alternatively, and preferably if the expected values are small, the log likelihood χ^2 test (G-test) can be used.

$$G = 2\Sigma \text{Obs} \ln \left(\frac{\text{Obs}}{\text{Exp}}\right)$$

$$= 2[\Sigma \text{Obs} \ln \text{Obs} - (2.30259)\Sigma \text{Obs} \log \text{Exp}]$$

The distribution of G will be approximated by the χ^2 distribution, and the significance of the value can be found in appropriate tables. (For further details of the G-test and its advantages see Sokal and Rohlf, 1969.)

An important point concerns the calculation of degrees of freedom when examining genotypic values. The number of independent genotypes is less than $N-1$ and the number of degrees of freedom can be calculated as

$$\tfrac{1}{2}(n^2-n)$$

where n is the number of alleles. Thus the 2-allele 3-genotype case has one degree of freedom, not 2 as in most $3\times 2\ \chi^2$ tests. Papers published as recently as 1976 contain incorrect degrees of freedom in their analyses.

Inter–population heterogeneity of allelic and genotypic frequencies

Inter-population heterogeneity in genotypic frequencies can be tested for using the *G*-test as above. *G*-values derived from independent comparisons may be added together, and the degrees of freedom corresponding to this are equal to the sum of the degrees of freedom for each set.

An examination for heterogeneity in allelic frequencies among populations can be carried out using the genic contingency chi-square test of Workman and Niswander (1970). If there are k alleles at a locus then the χ^2 statistic is given by

$$2N\left(\sum_{J=1}^{k}\frac{\sigma^2_{P_J}}{\bar{P}_J}\right)$$

where $\bar{P}_J$ and $\sigma^2_{P_J}$ are the weighted mean and variance of the frequencies of the Jth allele at the locus.

For each allele the weighted mean

$$\bar{p}=\Sigma(N_i/N)p_i$$

where p_i is the frequency of the allele in the ith population of sample size N_i and N is the total sample size of all populations being examined, i.e. instead of taking the mean frequency of an allele across all populations examined, it takes into account the sample sizes of each population.

The weighted variance $\sigma^2_p=\sum(N_i/N)p_i^2-\bar{p}^2$

The degrees of freedom are given by $(k-1)(r-1)$, where r is the number of populations examined.

Genetic identity and distance

Nei's coefficient of genetic identy (I) between two taxa is given by

$$I=\frac{\Sigma x_i y_i}{\sqrt{(\Sigma x_i^2\Sigma y_i^2)}}$$

where x_i and y_i are the frequencies of the ith allele in populations X and Y respectively

$$\text{(i.e. A, B, etc. to } n. \quad \Sigma x_i y_i = x_A y_A + x_B y_B \ldots + x_n y_n)$$

If in populations X and Y, the frequencies of alleles at a locus are as follows:

	A	B
Population X	0.46	0.54
Population Y	0.88	0.12

$$\text{then } I = \frac{(0.46 \times 0.88) + (0.54 \times 0.12)}{\sqrt{[(0.46^2 + 0.54^2) \times (0.88^2 + 0.12^2)]}}$$

$$= \frac{0.469}{0.63} = 0.745.$$

$I = 1$ when X and Y are monomorphic for the same allele and $I = 0$ when X and Y are monomorphic for different alleles.

The mean genetic identity ($\bar{I}$) is the mean over all loci studied (including monomorphic ones) and is most conveniently calculated as

$$\bar{I} = \frac{I_{XY}}{\sqrt{(I_X I_Y)}}$$

where I_{XY}, I_X and I_Y are the arithmetic means, over all loci of $\Sigma x_i y_i$, Σx_i^2 and Σy_i^2 respectively.

The genetic distance (D) is estimated by

$$D = -\ln I$$

which can be obtained from tables of natural logorithms or on many electronic calculators.

The time of divergence (T) of two taxa can be estimated by

$$T = 5 \times 10^6 \, D$$

(but see p. 74 and p. 103 for assumptions involved).

Presentation of data and construction of dendrograms

When genetic identity, genetic distance, or other measurements of genetic similarity have been compiled for all possible pairs of populations, species, etc. (hereafter referred to as operational taxonomic units or OTUs), these are best presented in the form of a matrix (table 6.3). For visual display of the results dendrograms can then be produced (p. 100). Various methods are available for dendrogram construction. The two most useful for electrophoretic data are the unweighted pair-group arithmetic average (UPGMA) clustering method, and the phylogenetic tree construction procedure of Fitch and Margoliash (1967). The latter method may give less distortion of the original data and does not assume uniform evolutionary rates (p. 100).

Unweighted pair-group arithmetic average (UPGMA) cluster analysis

In this method, the first two OTUs to be clustered are those with the highest genetic identity values (or lowest distance) between them. In table 10.2 the two most similar OTUs are 2 and 3. An appropriate scale covering the range of similarity values found in the OTUs to be clustered is drawn, and OTU 2 and OTU 3 are joined by a vertical line at an $\bar{I}$-value of 0.83 (figure 10.2). (The distance apart at which the OTUs are

Table 10.2 (data of Marinković *et al.*, 1978).

	OTU 2	OTU 3	OTU 4	OTU 5
OTU 1	0.48	0.36	0.35	0.27
OTU 2		0.83	0.12	0.03
OTU 3			0.05	0.01
OTU 4				0.53

Table 10.3

	OTU 2/3	OTU 4	OTU 5
OTU 1	0.42	0.35	0.27
OTU 2/3		0.09	0.02
OTU 4			0.53

Table 10.4

	OTU 2/3	OTU 4/5
OTU 1	0.42	0.31
OTU 2/3		0.05

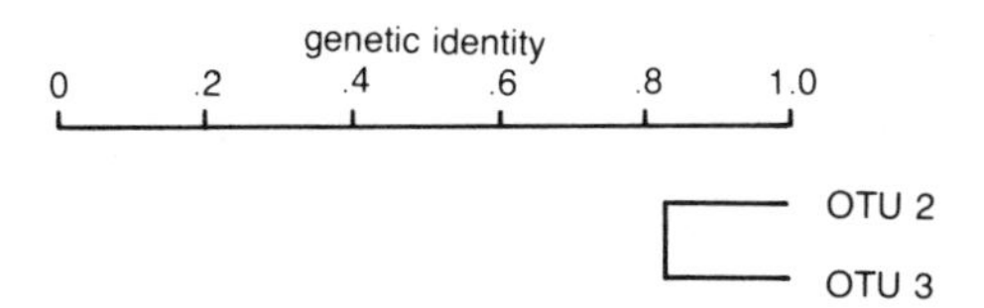

Figure 10.2. First stage in the construction of a dendrogram.

drawn is arbitrary.) The two OTUs are now combined as a single OTU 2/3, and a new matrix of values is calculated. The similarity value of OTU 2/3 to any other OTU is the mean of 2 to that OTU and the mean of 3 to that OTU. Thus the $\bar{I}$-value of OTU 2/3 to OTU 1 is

$$(0.48+0.36)/2 = 0.42.$$

The new matrix is given in table 10.3.

The next most similar pair is now joined in the dendrogram.

In the example this is OTU 4 and OTU 5. These two are joined by a line at an $\bar{I}$-value of 0.53 (figure 10.3).

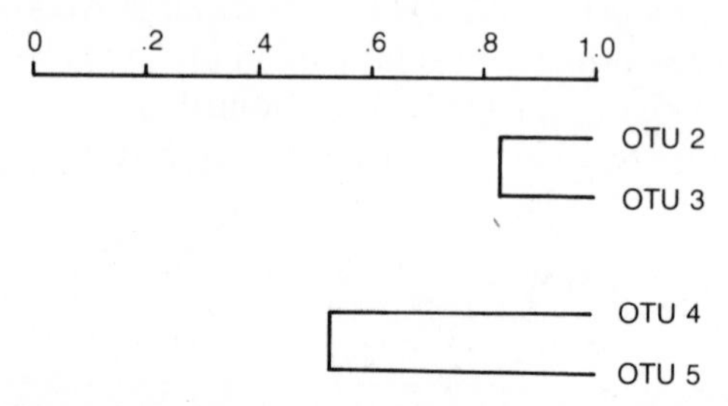

Figure 10.3. Second stage in dendrogram construction.

These now form a new group, OTU 4/5, and a matrix is recalculated as before (table 10.4). Again the similarity value of a grouped OTU is the mean of the similarities of its constituent members to that OTU.

The calculations are continued in cyclical fashion until all OTUs are part of the same cluster (figure 10.4).

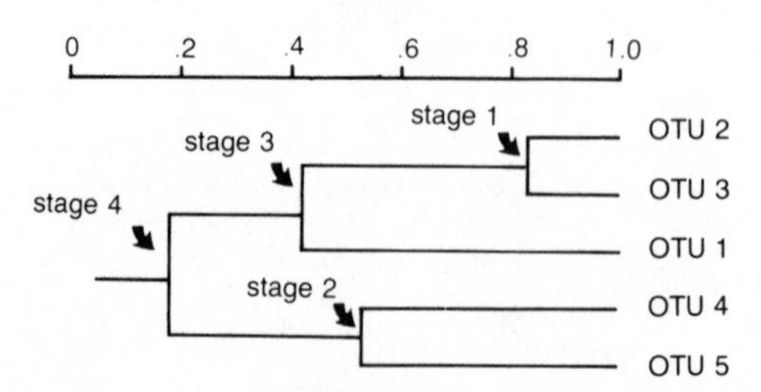

Figure 10.4 Stages in dendrogram construction.

Fitch and Margoliash method

Suppose three OTUs *A*, *B* and *C* have the following genetic distance values.

	B	*C*
A	.08	.19
B		.17

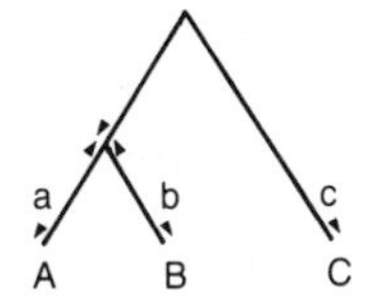

Figure 10.5. Initial phylogenetic tree of three OTUs.

A phylogenetic tree can be constructed by joining the pair of highest similarity, i.e. A and B, and then joining C to this group (figure 10.5). What then are the lengths of the branches a, b and c?

From the matrix it is known that

$$a+b = .08$$
$$a+c = .19$$
$$b+c = .17$$

Since $\quad 2a+b+c = .27 \quad \text{and} \quad b+c = .17,$

then $\quad 2a = .10 \quad \text{and} \quad a = .05$

Thus $b = .03$ and $c = .14$

Figure 10.6. Corrected phylogenetic tree of three OTUs.

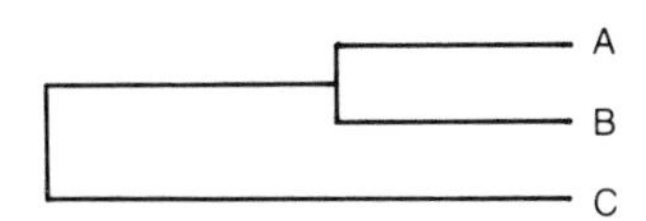

Figure 10.7. UPGMA dendrogram of same three OTUs as in figure 10.6.

A, B and C can now be joined by branches of appropriate lengths (figure 10.6). This tree illustrates the fact that more change has taken place in the line leading to A than to B, since they diverged from a common ancestor. Compare this with the UPGMA dendrogram (figure 10.7). The total tree length in figure 10.6 is .22 units, which exactly matches the input matrix.

If more than three OTUs are available, then to decide which two should be joined first, all possible pairwise combinations are used as A and B, with the remaining OTUs assigned to set C. Thus if there are six OTUs, then one is taken as A, one as B and the remaining four as C, and this is carried out for all fifteen possible alternative combinations. All OTUs are in each alternative part of one of three sets. The sets are then treated as for A, B and C above, except that now the similarity values are means of A to each member OTU of C, and likewise for B. The lowest distance from A to B is taken and these two are joined accordingly. Henceforth A and B are treated as a single OTU and the procedure repeated again. The number of OTUs is reduced by one after each cycle of calculations.

Since mean values are used in the calculation, the tree is unlikely to be an exact reconstruction of the input data. By adding together the distances between each pair of OTUs on the tree, it is possible to obtain a matrix of 'output' values. A comparison of the two matrices will give the difference between the input and output values. Other trees can then be constructed by joining a different initial pair (e.g. the second most similar) and a new tree constructed. The tree that gives the lowest deviation from the input matrix is accepted. This deviation can be conveniently estimated as the F-value of Prager and Wilson (1978), defined as $F = 100f/I$ where f is the sum of the absolute values of the differences between output and input values, and I is the sum of the input values.

Clearly the contruction of dendrograms and phylogenetic trees where many cyclical calculations are necessary can be carried out most efficiently on a computer or with the aid of a programmable calculator. Prager and Wilson (1978) state that, using a non-programmable calculator, a phylogenetic tree of twenty species was constructed by the Fitch-Margoliash method in seven hours, and required a further two hours to check.

APPENDIX

HINTS ON THE USE OF ELECTROPHORESIS IN SYSTEMATIC STUDIES

PRACTICAL DETAILS OF ELECTROPHORETIC TECHNIQUES ARE GIVEN BY Gordon (1975), Sargent and George (1975) and I. Smith (1976). Of these, Smith is probably the most useful, as both horizontal and vertical starch gel electrophoresis are covered, and details of enzyme staining methods are given. A good short account is that of Oelshlegel and Stahmann (1973). Books devoted specifically to enzyme staining techniques are Brewer (1970) and Harris and Hopkinson (1976 and supplements). The latter is an excellent laboratory handbook, providing methods for staining some 80 different enzymes. Although primarily developed for human studies, many of the methods can be used without alteration on other organisms. In addition, it provides background information on the staining reactions involved, subunit structure and tissue distribution.

Details which can be found in Smith, or in Harris and Hopkinson, are not repeated here; only those additions and amendments which the author has found to be of value in the application of electrophoresis to systematic problems in a wide range of animals are given.

Collection of samples

In most cases, proteins have to be extracted from tissues. This is conveniently carried out by homogenization of the tissue with an equal volume of buffer (e.g. 1.21 g Tris(hydroxymethyl)aminomethane and 0.37 g Na_2 EDTA per litre, adjusted to pH 7.0 with dilute HCl). When dealing with a large number of samples, a suitable system can be built around 5-ml plastic screw-cap disposable tubes (e.g. Monoject[1] C/5/SC). Up to 2 ml of tissue and buffer can be homogenized in this tube, using a glass rod or a motor-driven homogenizer (e.g. I.L.A. homogenizer with

10-mm shaft, Scientific Instrument Centre Ltd[2].). These tubes have the further advantage that they fit the standard 17-mm centrifuge holders and thus centrifugation can be carried out without the need for homogenate to be transferred to separate centrifuge tubes. After centrifugation, the sample can be stored in the same tube. Filing racks are available for freezer use, and these make the location of individual samples much easier (Luckham[3], Denley[4]).

Starch gel preparation

Starch gels are prepared by heating partially hydrolysed starch in a suitable gel buffer. Two main types of starch are commercially available: Connaught (available from most biochemical suppliers) and Electrostarch[5]. Electrostarch produces a gel with greater mechanical strength, which makes subsequent handling of the gel much easier. The two types differ in their separation properties, as do individual batches of the same type.

A large variety of gel and electrode-tank buffers has been used for electrophoresis of proteins. The following buffer system is a good general one for initial screening purposes:

Gel buffer. 9.2 g Tris, 1.05 g citric acid and 53 ml electrode buffer per litre.
Electrode buffer. 18.6 g boric acid and 4.2 g lithium hydroxide per litre.
Both gel and electrode buffers have a pH of 8.6

For some enzymes (e.g. malate dehydrogenase, glycerol-3-phosphate dehydrogenase, glutamate oxaloacetate transaminase), superior resolution and staining intensity is obtained when an electrophoresis buffer of pH 6.0–7.0 is used. The following low pH buffers are suitable:

N-(3-Aminopropyl)-morpholine–citrate (Clayton and Tretiak, 1972).
Gel–0.002 M citric acid, pH 6.0.
Electrode–0.04 M citric acid, pH 6.1.
Buffers are adjusted to correct pH with N-(3-Aminopropyl)-morpholine (Aldrich Chemical Co.).
Histidine–citrate (Brewer, 1970)
Gel–0.005 M histidine, adjusted to pH 7.0 with 2N sodium hydroxide
Electrode–0.41 M sodium citrate adjusted to pH 7.0 with 0.41 M citric acid.

Details of gel preparation are given in chapter 7 of Smith. The rate of heating, the final temperature and the time of de-aeration should be standardized to ensure uniform gels. A final temperature of 80°C and de-aeration for 1 minute at a vacuum of 1 bar are suggested. Heating of the starch-buffer suspension can be carried out conveniently in a saucepan, thus allowing adequate stirring, rather than in a side-arm flask as Smith suggests. The hot starch is then poured into a side-arm flask of at least twice its volume, for de-aeration.

Gel moulds and tanks for horizontal gel electrophoresis are constructed easily from Perspex. A mould of 180 mm × 160 mm × 6 mm (internal dimensions) is a suitable size and allows 20–25 samples to be screened simultaneously. Samples are applied to the horizontal gels on 5-mm squares of Whatman No. 3 filter paper, which are inserted in a slit in the gel situated about 30 mm from the cathodal end of the gel for pH 8.6 systems, or in the middle of the gel for lower pHs. The slit is then closed by means of a Perspex compressor strip (155 mm × 10 mm × 6 mm) which is inserted at the anodal end of the gel. The completed set up is shown in figure 3.3.

Apparatus for vertical starch gel electrophoresis is available from Buchler Instruments[6]. In the case of vertical gels the sample is pipetted into a slot in the gel, which is then sealed with molten Vaseline.

A direct-current power supply capable of producing some 300 V at 100 mA is required. This should preferably be stabilized for consistent results. One capable of supplying constant power is ideal for optimum speed of separation.

After electrophoresis, a 6-mm-thick gel can be cut into 2–4 slices which can be stained independently. The stain should be applied to the cut surface of the gel, and not to the top or bottom faces. Plastic lunch boxes are convenient staining containers.

Acrylamide gels

Details of acrylamide gel preparation are given in the above references. Ready-for-use gradient pore gels (Universal Scientific[7]) and isolectric focusing gels of various pH ranges (LKB[8]) are commercially available.

Addresses of suppliers

1. Monoject Division, Sherwood Medical, London Road, County Oak, Crawley, West Sussex, RH10 2TL.
2. Scientific Instrument Centre Ltd., 13 Derby Lane, Liverpool L13 6QA.
3. Luckham Ltd., Labro Works, Victoria Gardens, Burgess Hill, Sussex.
4. Denley Instruments Ltd., Billingshurst, Sussex.
5. Electrostarch Co., P.O. Box 1294, Madison, Wisconsin 5370, U.S.A.

   or

   Strand Scientific, 32 Bridge Street, Sandiacre, Nottingham.
6. Available in U.K. from Arnold Horwell Ltd., 2 Grangeway, Kilburn High Road, London NW6 2BP.
7. Universal Scientific Ltd., 231 Plashet Road, London E13.
8. LKB Instruments Ltd., 232 Addington Road, South Croydon, Surrey CR2 8YD.

FURTHER READING

General works

Ayala, F. J. (1976) *Molecular Evolution*, Sinauer Associates, Inc., Massachusetts.

Dobzhansky, T., Ayala, F. J., Stebbins, G. L. and Valentine, J. W. (1977) *Evolution*, Freeman and Co., San Francisco.

Grant, V. (1977) *Organismic Evolution*, Freeman and Co., San Francisco.

Manwell, C. and Baker, C. M. A. (1970) *Molecular Biology and the Origin of Species*, Sidgwick and Jackson, London.

Nei, M. (1975) *Molecular Population Genetics and Evolution*, North-Holland, Amsterdam.

Smith, P. M. (1976) *The Chemotaxonomy of Plants*, Edward Arnold, London.

Wright, C. A. (Ed.) (1974) *Biochemical and Immunological Taxonomy of Animals*, Academic Press, London.

Chapter 1

Hennig, W. (1966) *Phylogenetic Systematics*, University of Illinois Press, Urbana.

Mayr, E. (1969) *Principles of Systematic Zoology*, McGraw-Hill Book Company, New York.

Mayr, E. (1970) *Populations, Species, and Evolution*, Belknap Press, Massachusetts.

Mayr, E. (1974) 'Cladistic Analysis or Cladistic Classification', *Z. zool. Syst. Evolut.-forsch.* **12**, 94–128.

Sneath, P. H. A. and Sokal, R. R. (1973) *Numerical Taxonomy*, Freeman and Co., San Francisco.

White, M. J. D. (1978) *Modes of Speciation*, Freeman and Co., San Francisco.

Chapter 2

Dickerson, R. E. D. and Geis, I. (1969) *The Structure and Action of Proteins*, W. A. Benjamin, Inc., London.

Eigen, M. and Winkler-Oswatitsch, R. (1976) 'The Game of Evolution', *Interdisciplinary Science Reviews* **1**, 19–30.

MacIntyre, R. J. (1976) 'Evolution and Ecological Value of Duplicate Genes', *Ann. Rev. Ecol. Syst.* **7**, 421–468.

Chapter 3

Catsimpoolas, N. (Ed.) (1975–76) *Methods of Protein Separation*, Vols. 1 and 2, Plenum Press, London.

Champion, A. B., Prager, E. M., Wachter, D. and Wilson, A. C. (1974) 'Microcomplement Fixation', *Biochemical and Immunological Taxonomy* of Animals, Ed. Wright, C. A., Academic Press, London.

Kohne, D. E. (1970) 'Evolution of Higher-organism DNA', *Quart. Rev. of Biophys.* **3**, 327–365.

Konigsberg, W. H. and Steinman, H. W. (1977) 'Strategy and Methods of Sequence Analysis', *The Proteins*, Vol. III, Ed. Neurath, H. and Hill, R. L., Academic Press, London.

(See Appendix for further references to electrophoretic techniques).

Chapter 4

Berry, R. J. (1977) *Inheritance and Natural History*, Collins, London.

Christiansen, F. B. and Fenchel, T. M. (1977) *Measuring Selection in Natural Populations*, Lecture Notes in Biomathematics No. 19, Springer-Verlag, Berlin.

Harris, H. (1976) 'Molecular Evolution: the Neutralist–Selectionist Controversy', *Fed. Proc.* **35**, 2079–2082.

Lewontin, R. C. (1974) *The Genetic Basis of Evolutionary Change*, Columbia University Press, New York.

Lucotte, G. (1977) *Le Polymorphisme Biochimique et les Facteurs de son Maintien*, Masson, Paris.

Markert, C. L. (1975) *Isozymes*, Vols. 1, 2, 3 and 4, Academic Press, London.

Chapter 5

Godfrey, D. G. (1978) 'Identification of Economically Important Parasites', *Nature, Lond.* **273**, 600–604.

Hedgecock, D., Shleser, R. A. and Nelson, K. (1976) 'Applications of Biochemical Genetics to Aquaculture', *J. Fish. Res. Board Can.* **33**, 1108–1119.

Jamieson, A. (1974) 'Genetic Tags for Marine Fish Stocks', *Sea Fisheries Research*, Ed. Harden Jones, F. R., Elek Science, London.

Moav, R., Brody, T., Wohlfarth, G. and Hulata, G. (1976) 'Applications of Electrophoretic Genetic Markers to Fish Breeding. 1. Advantages and Methods', *Aquaculture* **9**, 217–228.

Utter, F. M., Hodgins, H. O. and Allendorf, F. W. (1974) 'Biochemical Genetic Studies of Fishes: Potentialities and Limitations', *Biochemical and Biophysical Perspectives in Marine Biology*, Ed. Malins, D. C. and Sargent, J. R., Academic Press, London.

Utter, F. M., Allendorf, F. W. and May, B. (1976) 'The Use of Protein Variation in the Management of Salmonid Populations', *Trans. North Am. Wildl. Nat. Resour.* **41**, 373–384.

Chapter 6

Avise, J. C. (1975) 'Systematic Value of Electrophoretic Data', *Syst. Zool.* **23**, 465–481.

Avise, J. C. and Ayala, F. J. (1976) 'Genetic Differentiation in Speciose versus Depauperate Phylads: Evidence from the California Minnows', *Evolution* **30**, 46–58.

Avise, J. C. and Smith, M. H. (1977) 'Gene Frequency Comparisons between Sunfish (Centrarchidae) Populations at Various Stages of Evolutionary Divergence', *Syst. Zool.* **26**, 319–335.

Ayala, F. J. (1975) 'Genetic Differentiation during the Speciation Process', *Evolutionary Biology*, Vol. 8, Ed. Dobzhansky, T., Hecht, M. K. and Steere, W. C., Plenum Press, New York.

Sarich, V. M. (1977) 'Rates, Sample Sizes, and the Neutrality Hypothesis for Electrophoresis in Evolutionary Studies', *Nature, Lond.* **265**, 24–28.

Schechter, Y. (1973) 'Symposium on the Use of Electrophoresis in the Taxonomy of Algae and Fungi', *Bull. Torrey Bot. Club* **100**, 253–259.

Throckmorton, L. H. (1977) '*Drosophila* Systematics and Biochemical Evolution', *Ann. Rev. Ecol. Syst.* **8**, 235–254.

Chapter 7

Jackson, R. C. (1976) 'Evolution and Systematic Significance of Polyploidy', *Ann. Rev. Ecol. Syst.* **7**, 209–234.

Kung, S-D. (1976) 'Tobacco Fraction 1 Protein: A Unique Genetic Marker', *Science* **191**, 429–434.

Chapter 8

De Haën, C. and Neurath, H. (1976) 'Protein Structure and the Evolution of Species', *Biochemical and Biophysical Perspectives in Marine Biology*, Vol. 3, Academic Press, London.

Fitch, W. M. (1976) 'The Molecular Evolution of Cytochrome c in Eukaryotes', *J. Mol. Evol.* **8**, 13–40.

Fitch, W. M. (1977) 'The Phyletic Interpretation of Macromolecular Sequence Information', *Major Patterns in Vertebrate Evolution*, Eds. Hecht, W. M., Goody, P. C. and Hecht, B. M., Plenum Press, London.

Goodman, M. and Tashian, R. E. (Eds) (1976) *Molecular Anthropology*, Plenum Press, London.

Kohne, D. E. (1970) 'Evolution of Higher-organism DNA', *Quart. Rev. of Biophys.* **3**, 327–375.

Wilson, A. C., Carlson, S. S. and White, T. J. (1977) 'Biochemical Evolution', *Ann. Rev. Biochem.* **46**, 573–639.

Chapter 9

King, M. C. and Wilson, A. C. (1975) 'Evolution at Two Levels in Humans and Chimpanzees'. *Science* **188**, 107–116.

Wilson, A. C. (1976) 'Gene Regulation in Evolution', *Molecular Evolution*, Ed. Ayala, F. J., Sinauer Associates Inc., Massachusetts.

Source of further references

The Savannah River Ecology Laboratory (Drawer E. Aiken, South Carolina 29801, U.S.A.) produces a cross-indexed bibliography of electrophoretic studies of natural populations of vertebrates. The first edition, published in 1979, contains 500+ references and is available as a printout or computer tape.

COMPREHENSIVE BIBLIOGRAPHY

Abramoff, P., Darnell, R. M. and Balsano, J. S. (1968) 'Electrophoretic Demonstration of the Hybrid Origin of the Gynogenetic Teleost *Poecilia formosa*', *Amer. Natur.* **102**, 555–558.

Adams, J. and Allen, S. (1975) 'Genetic Polymorphism and Differentiation in *Paramecium*', in *Isozymes*, Vol. IV, Ed. Markert, C. L., Academic Press, London.

Allard, R. W. and Kahler, A. L. (1972) 'Patterns of Molecular Variation in Plant Populations', in *Proceedings of the Sixth Berkeley Symposium on Mathematical Statistics and Probability*, Vol. 5, Ed. Le Cam, L. M., Neyman, J. and Scott, E. L., University of California Press, Berkeley.

Allen, S. and Gibson, I. (1975) 'Syngenic Variations for Enzymes of *Paramecium aurelia*', in *Isozymes*, Vol. IV, Ed. Markert, C. L., Academic Press, London.

Allendorf, F., Ryman, N., Stennek, A. and Ståhl, G. (1976) 'Genetic Variation in Scandinavian Brown Trout (*Salmo trutta* L.): Evidence of Distinct Sympatric Populations', *Hereditas* **83**, 73–82.

Allendorf, F. M., Utter, F. M. and May, B. P. (1975) 'Gene Duplication within the Family Salmonidae: 11. Detection and Determination of the Genetic Control of Duplicate Loci through Inheritance Studies and the Examination of Populations', in *Isozymes*, Vol. IV, Ed. Markert, C. L., Academic Press, London.

Almgård, G. and Landegren, U. (1974) 'Isoenzymatic Variation used for the Identification of Barley Cultivars', *Z. Pflanzenzüchtg.***72**, 63–73.

Anderson, L. and Anderson, N. G. (1977) 'High Resolution Two-dimensional Electrophoresis of Human Plasma Proteins', *Proc. Natl. Acad. Sci. USA* **74**, 5421–5425.

Ashton, G. C., Fallon, G. R. and Sutherland, D. N. (1964) 'Transferrin (β-globulin) Type and Milk and Butterfat Production in Dairy Cows', *J. agric. Sci.* **62**, 27–34.

Aspinwall, N. (1974) 'Genetic Analysis of North American Populations of the Pink Salmon, *Oncorhynchus gorbuscha*, Possible Evidence for the Neutral Mutation–Random Drift Hypothesis', *Evolution* **28**, 295–305.

Avise, J. C. (1975) 'Systematic Value of Electrophoretic Data', *Syst. Zool.* **23**, 465–481.

Avise, J. C. and Ayala, F. J. (1976) 'Genetic Differentiation in Speciose versus Depauperate Phylads: Evidence from the California Minnows', *Evolution* **30**, 46–58.

Avise, J. C. and Duvall, S. W. (1977) 'Allelic Expression and Genetic Distance in Hybrid Macaque Monkeys', *J. Hered.* **68**, 23–30.

Avise, J. C. and Smith, M. H. (1977) 'Gene Frequency Comparisons between Sunfish (Centrarchidae) Populations at Various Stages of Evolutionary Divergence', *Syst. Zool.* **26**, 319–335.

Avise, J. C., Straney, D. O. and Smith, M. H. (1977) 'Biochemical Genetics of Sunfish IV. Relationship of Centrarchid Genera', *Copeia* **1977**, 250–258.

Ayala, F. J. (1975) 'Genetic Differentiation during the Speciation Process', in *Evolutionary Biology*, Vol. 8, Ed. Dobzhansky, T., Hecht, M. and Steere, W. C., Plenum Press, New York.

Ayala, F. J., Tracey, M. L., Barr, L. G., McDonald, J. F. and Pérez-Salas, S. (1974) 'Genetic Variation in Natural Populations of Five *Drosophila* Species and the Hypothesis of the Selective Neutrality of Protein Polymorphisms', *Genetics* **77**, 343–384.

Bagster, I. A. and Parr, C. W. (1973) 'Trypanosome Identification by Electrophoresis of Soluble Enzymes', *Nature* **244**, 364–366.

Bakay, B., Nyham, W. L. and Monkus, E. St. J. (1972) 'Change in Electrophoretic Mobility of Glucose-6-phosphate Dehydrogenase with Ageing of Erythrocytes', *Pediat. Res.* **6**, 705–712.

Baker, C. M. A. and Manwell, C. (1975) 'Molecular Biology of Avian Proteins–XII. Protein Polymorphism in the Stubble Quail *Coturnix pectoralis*–and a Brief Note on the Induction of Egg White Protein Synthesis in Wild Birds by Hormones', *Comp. Biochem. Physiol.* **50B**, 471–477.

Baker, C. M. A., Manwell, C., Labisky, R. F. and Harper, J. A. (1966) 'Molecular Genetics of Avian Proteins–V. Egg, Blood and Tissue Proteins of the Ring-necked Pheasant, *Phasianus colchicus* L.', *Comp. Biochem. Physiol.* **17**, 467–499.

Baldwin, T. O. and Riggs, A. (1974) 'The Haemoglobins of the Bullfrog, *Rana catesbiana*. Partial Amino Acid Sequence of the β Chain of the Major Adult Component', *J. Biol. Chem.* **249**, 6110–6118.

Balsano, J. S., Darnell, R. N. and Abramoff, P. (1972) 'Electrophoretic Evidence of Triploidy Associated with Populations of the Gynogenetic Teleost *Poecilia formosa*', *Copeia* **1972**, 292–297.

Bateson, W. (1913) *Problems of Genetics*, Oxford Univ. Press, London.

Bergman, H., Carlstrom, A., Gustavsson, I. and Lindsten, J. (1971) 'Protein-Heparin Complexes as a Cause of Artificial Enzyme Polymorphism', *Scand. J. Clin. Lab. Invest.* **27**, 341–344.

Berrie, A. D. (1973) 'Snails, Schistosomes and Systematics: Some Problems Concerning the Genus *Bulinus*', in *Taxonomy and Ecology*, Ed. Heywood, V. H., Academic Press, London.

Berry, R. J. (1977) *Inheritance and Natural History*, Collins, London.

Beyer, W. A., Stein, M. L., Smith, T. F. and Ulam, S. M. (1974) 'A Molecular Sequence Metric and Evolutionary Trees', *Mathl. Biosciences* **19**, 9–25.

Bisbee, C. A., Baker, M. A., Wilson, A. C., Hadji-Azimi, I. and Fischberg, M. (1977) 'Albumin Phylogeny for Clawed Frogs (*Xenopus*)', *Science* **195**, 785–787.

Blackwelder, R. E. (1967) *Taxonomy*, Wiley, New York.

Bodmer, W. F. and Cavalli-Sforza, L. L. (1976) *Genetics, Evolution, and Man*, W. H. Freeman & Co., San Francisco.

Bogart, J. P. and Tandy, M. (1976) 'Polyploid Amphibians: Three more Diploid-Tetraploid Cryptic Species of Frogs', *Science* **193**, 334–337.

Bonnell, M. L. and Selander, R. K. (1974) 'Elephant Seals: Genetic Variation and near Extinction', *Science* **184**, 908–909.

Borden, D. Miller, E. T., Whitt, G. S. and Nanney, D. L. (1977) 'Electrophoretic Analysis of Evolutionary Relationships in *Tetrahymena*', *Evolution* **31**, 91–102.

Brassington, R. A. and Ferguson, A. (1976) 'Electrophoretic Identification of Roach (*Rutilus rutilus* L.), Rudd (*Scardinius erythrophthalmus* L.), Bream (*Abramis brama* L.) and their Natural Hybrids', *J. Fish Biol.* **9**, 471–477.

Brewer, G. J. (1970) *An Introduction to Isozyme Techniques*, Academic Press, London.

Brewer, J. M. (1967) 'Artifacts Produced in Disc Electrophoresis by Ammonium Persulfate', *Science* **156**, 256–257.

Britten, R. J. and Davidson, E. H. (1976) 'DNA Sequence Arrangement and Preliminary Evidence on its Evolution', *Fedn. Proc.* **35**, 2151–2157.

Brittnacher, J. C., Sims, S. R. and Ayala, F. J. (1978) 'Genetic Differentiation between Species of the Genus *Speyeria* (Lepidoptera: Nymphalidae)', *Evolution* **32**, 199–210.

Britton, J. and Thaler, L. (1978) 'Evidence for the Presence of Two Sympatric Species of Mice (Genus *Mus* L.) in Southern France Based on Biochemical Genetics', *Biochem. Genet.* **16**, 213–225.

Brush, A. H. (1976) 'Waterfowl Feather Proteins: Analysis of use in Taxonomic Studies', *J. Zool., Lond.* **179**, 467–498.

Bryce, D. and Hobart, A. (1972) 'The Biology and Identification of the Larvae of the Chironomidae (Diptera)', *Entomologist's Gazette* **23**, 175–217.

Burns, J. M. (1975) 'Isozymes in Evolutionary Systematics' in *Isozymes*, Vol. IV, Ed. Markert, C. L., Academic Press, London.

Bylund, G. and Djupsund, B. M. (1977) 'Protein Profiles as an Aid to Taxonomy in the Genus *Diphyllobothrium*', *Z. Parasitenk.* **51**, 241–247.

Cann, J. R. (1966) 'Multiple Electrophoretic Zones Arising from Protein-buffer Interaction', *Biochemistry* **5**, 1108–1112.

Cavalli-Sforza, L. L. (1969) 'Human Diversity', *Proc. XII Intern. Congr. Genet.* **3**, 405–416.

Chai, C. K. (1976) *Genetic Evolution*, University of Chicago Press, London.

Champion, A. B., Prager, E. M., Wachter, D. and Wilson, A. C. (1974) 'Microcomplement Fixation' in *Biochemical and Immunological Taxonomy of Animals*, Ed. Wright, C. A., Academic Press, London.

Champion, A. B., Soderberg, K. L., Wilson, A. C. and Ambler, R. P. (1975) 'Immunological Comparison of Azurins of Known Amino Acid Sequence', *J. Mol. Evol.* **5**, 291–305.

Chen, K., Gray, J. C. and Wildman, S. G. (1975) 'Fraction 1 Protein and the Origin of Polyploid Wheats', *Science* **190**, 1304–1306.

Child, A. R., Burnell, A. M. and Wilkins, N. P. (1976) 'The Existence of Two Races of Atlantic Salmon (*Salmo salar* L.) in the British Isles', *J. Fish Biol.* **8**, 35–44.

Child, A. R. and Solomon, D. J. (1977) 'Observations on Morphological and Biochemical Features of Some Cyprinid Hybrids', *J. Fish Biol.* **11**, 125–131.

Christiansen, F. B. (1977) 'Genetics of *Zoarces* populations X. Selection Component Analysis of the Est 111 Polymorphism Using Samples of Successive Cohorts', *Hereditas* **87**, 129–150.

Clayton, J. W. and Tretiak, D. N. (1972) 'Amine-citrate Buffers for pH Control in Starch Gel Electrophoresis', *J. Fish. Res. Board Can.* **29**, 1169–1172.

Clegg, M. T. and Allard, R. W. (1972) 'Patterns of Genetic Differentiation in the Slender Wild Oat Species *Avena barbata*', *Proc. natn. Acad. Sci. U.S.A.* **69**, 1820–1824.

Collier, G. E. and MacIntyre, R. J. (1977) 'Microcomplement Fixation Studies on the Evolution of α-Glycerophosphate Dehydrogenase within the Genus Drosophila', *Proc. natn. Acad. Sci. U.S.A.* **74**, 684–688.

Coluzzi, M. and Bullini, L. (1971) 'Enzymatic Variants as Markers in the Study of Precopulatory Isolating Mechanisms', *Nature* **231**, 455–456.

Cooper, D. W., Johnson, P. G., Murtagh, C. E., Sharman, G. B., Vandeberg, J. L. and Poole, W. E. (1975) 'Sex-linked Isozymes and Sex Chromosome Evolution and Interaction in Kangaroos', in *Isozymes* Vol. III, Ed. Markert, C. L., Academic Press, London.

Cross, T. F. and Payne, R. H. (1978) 'Geographic Variation in Atlantic Cod, *Gadus morhua*, off Eastern North America: A Biochemical Systematics Approach', *J. Fish. Res. Board Can.* **35**, 117–123.

Crozier, R. H. (1973) 'Apparent Differential Selection at an Isozyme Locus between Queens and Workers of the Ant *Aphaenogaster rudis*', *Genetics* **73**, 313–318.

Day, T. H., Hillier, P. C. and Clarke, B. (1974) 'Properties of Genetically Polymorphic Isozymes of Alcohol Dehydrogenase in *Drosophila melanogaster*', *Biochem. Genet.* **11**, 141–153.

Dayhoff, M. O. (1972) *Atlas of Protein Sequence and Structure*, Vol. 5, Natl. Biomed. Res. Foundation, Washington.

Dayhoff, M. O. (1973) *Atlas of Protein Sequence and Structure*, Vol. 5, Supplement 1, Natl. Biomed. Res. Foundation, Washington.

Dayhoff, M. O. (1976) *Atlas of Protein Sequence and Structure*, Vol. 5, Supplement 2, Natl. Biomed. Res. Foundation, Washington.

Dessauer, H. C. and Nevo, E. (1969) 'Geographic Variation of Blood and Liver Proteins in Cricket Frogs', *Biochem. Genet.* **3**, 171–188.

Dickerson, R. E. D. and Geis, I. (1969) *The Structure and Action of Proteins*, Benjamin, London.

Doane, W. W., Abraham, I., Kolar, M. M., Martenson, E. and Deibler, G. E. (1975) 'Purified *Drosophila* α-Amylase Isozymes: Genetical, Biochemical, and Molecular Characterization', in *Isozymes* Vol. IV, Ed. Markert, C. L., Academic Press, London.

Dobzhansky, T. (1973) 'Nothing in Biology Makes Sense Except in the Light of Evolution', *Amer. Biology Teacher* **35**, 125–129.

Dobzhansky, T., Ayala, F. J., Stebbins, G. L. and Valentine, J. W. (1977) *Evolution*, W. H. Freeman & Co., San Francisco.

Eriksson, K., Halkka, O., Lokki, J. and Saura, A. (1976) 'Enzyme Polymorphism in Feral, Outbred and Inbred Rats (*Rattus norvegicus*)', *Heredity, Lond.* **37**, 341–349.

Everhart, W. H., Eipper, A. W. and Youngs, W. D. (1975) *Principles of Fishery Science*, Cornell University Press, Ithaca.

Fantes, K. H. and Furminger, I. G. S. (1967) 'Proteins, Persulfate and Disc Electrophoresis', *Nature* **215**, 750–751.

Farris, J. S. (1972) 'Estimating Phylogenetic Trees from Distance Matrices', *Amer. Nat.* **106**, 645–668.

Fedak, G. (1974) 'Allozymes as Aids to Canadian Barley Cultivar Identification', *Euphytica* **23**, 166–173.

Ferguson, A. (1974) 'The Genetic Relationships of the Coregonid Fishes of Britain and Ireland Indicated by Electrophoretic Analysis of Tissue Proteins', *J. Fish Biol.* **6**, 311–315.

Ferguson, A. (1975) 'Myoglobin Polymorphism in the Pollan (Osteichthyes: Coregoninae)' *Amin. Blood Grps biochem. Genet.* **6**, 25–29.

Ferguson, A., Himberg, K-J. M. and Svärdson, G. (1978) 'Systematics of the Irish Pollan (*Coregonus pollan* Thompson): An Electrophoretic Comparison with Other Holarctic Coregoninae', *J. Fish Biol.* **12**, 221–233.

Ferris, S. D. and Whitt, G. S. (1977*a*) 'Loss of Duplicate Gene Expression after Polyploidisation', *Nature* **265**, 258–260.

Ferris, S. D. and Whitt, G. S. (1977*b*) 'Duplicate Gene Expression in Diploid and Tetraploid Loaches (Cypriniformes, Cobitidae)', *Biochem. Genet.* **15**, 1097–1112.

Ferris, S. D. and Whitt, G. S. (1977*c*) 'The Evolution of Duplicate Gene Expression in the Carp (*Cyprinus carpio*)', *Experientia* **33**, 1299–1301.

Ferris, S. D. and Whitt, G. S. (1978) 'Phylogeny of Tetraploid Catostomid Fishes Based on the Loss of Duplicate Gene Expression', *Syst. Zool.* **27**, 189–206.

Fitch, W. M. (1976) 'Molecular Evolutionary Clocks', in *Molecular Evolution*, Ed. Ayala, F. J., Sinauer Associates, Sunderland, Massachusetts.

Fitch, W. M. (1977) 'The Phyletic Interpretation of Macromolecular Sequence Information', in *Major Patterns in Vertebrate Evolution*, Ed. Hecht, M. K., Goody, P. C. and Hecht, B. M., Plenum Press, London.

Fitch, W. M. and Margoliash, E. (1967) 'Construction of Phylogenetic Trees', *Science* **155**, 279–284.

Flake, R. H. and Lennington, R. K. (1977) 'Genetic Distances Utilizing Electrophoretic Mobility Data', *Biol. Zbl.* **96**, 451–456.

Fowle, K. E., Cline, J. H., Klosterman, E. W. and Parker, C. F. (1967) 'Transferrin Genotypes and their Relationship with Blood Constituents, Fertility and Cow Productivity', *J. Anim. Sci.* **26**, 1226–1231.

Frelinger, J. A. (1972) 'The Maintenance of Transferrin Polymorphism in Pigeons', *Proc. natn. Acad. Sci. U.S.A.* **69**, 326–329.

Frelinger, J. A. (1973) 'Chemical Basis of Transferrin Polymorphism in Pigeons', *Anim. Blood Grps. biochem. Genet.* **4**, 35–40.

Frydenberg, O. and Simonsen, V, (1973) 'Genetics of *Zoarces* Populations V. Amount of Protein Polymorphism and Degree of Genic Heterozygosity', *Hereditas* **75**, 221–232.

Fujino, K. and Kang, T. (1968) 'Transferrin Groups of Tunas', *Genetics* **59**, 79–91.

Garten, C. T. (1977) 'Relationships between Exploratory Behaviour and Genic Heterozygosity in the Oldfield Mouse', *Anim. Behav.* **25**, 328–332.

Gibson, W. C., Parr, C. W., Swindlehurst, C. A. and Welch, S. G. (1978) 'A Comparison of the Isoenzymes, Soluble Proteins, Polypeptides and Free Amino Acids from Ten Isolates of *Trypanosoma evansi*', *Comp. Biochem. Physiol.* **60B**, 137–142.

Gilmour, D. G. and Morton, J. R. (1971) 'Association of Genetic Polymorphisms with Embryonic Mortality in the Chicken III. Interactions between Three Loci Determining Egg-white Proteins', *Theor. Appl. Genet.* **41**, 57–66.

Goodman, M. (1976) 'Protein Sequences in Phylogeny', in *Molecular Evolution*, Ed. Ayala, F. J., Sinauer Associates, Sunderland, Massachusetts.

Goodman, M. and Tashian, R. E. (1976) *Molecular Anthropology*, Plenum Press, London.

Gordon, A. H. (1975) *Electrophoresis of Proteins in Polyacrylamide and Starch Gels*, North-Holland, Amsterdam.

Gottlieb, L. D. (1976) 'Biochemical Consequences of Speciation in Plants', in *Molecular Evolution*, Sinauer Associates, Sunderland, Massachusetts.

Grant, V. (1977) *Organismic Evolution*, W. H. Freeman & Co., San Francisco.

Gray, J. C., Kung, S. D., Wildman, S. G. and Sheen, S. J. (1974) 'Origin of *Nicotiana tabacum* L. Detected by Polypeptide Composition of Fraction 1 Protein', *Nature* **252**, 226–227.

Guttman, S. I. (1975) 'Genetic Variation in the Genus *Bufo* II. Isozymes in Northern Allopatric Populations of the American Toad *Bufo Americanus*', in *Isozymes* Vol. IV, Ed. Markert, C. L., Academic Press, London.

Hare, D. L., Stimpson, D. T. and Cann, J. R. (1978) 'Multiple Bands Produced by Interaction of a Single Macromolecule with Carrier Ampholytes during Isoelectric Focusing', *Archs. Biochem. Biophys.* **187**, 274–275.

Harris, H. and Hopkinson, D. A. (1972) 'Average Heterozygosity per Locus in Man: An Estimate Based on the Incidence of Enzyme Polymorphism', *Ann. Hum. Genet.* **36**, 9–20.

Harris, H. and Hopkinson, D. A. (1976) *Handbook of Enzyme Electrophoresis in Human Genetics*, North-Holland, Amsterdam.

Harrison, R. G. (1977) 'Parallel Variation at an Enzyme Locus in Sibling Species of Field Crickets', *Nature* **266**, 168–170.

Haslett, B. G., Evans, I. M. and Boulter, D. (1978) 'Amino Acid Sequence of Plastocyanin from *Solanum crispum* using Automatic Methods', *Phytochemistry* **17**, 735–739.

Hedgecock, D., Shleser, R. A. and Nelson, K. (1976) 'Application of Biochemical Genetics to Aquaculture', *J. Fish. Res. Board Can.* **33**, 1108–1119.

Hedrick, P. W. (1971) 'A New Approach to Measuring Genetic Similarity', *Evolution* **25**, 276–280.

Hennig, W. (1966) *Phylogenetic Systematics*, Univ. of Illinois Press, Urbana.

Holmquist, R. (1976) 'Random and Nonrandom Processes in the Molecular Evolution of Higher Organisms', in *Molecular Anthropology*, Ed. Goodman, M. and Tashian, R. E., Plenum Press, London.

Hopkinson, D. A., Edwards, Y. H. and Harris, H. (1976) 'The Distributions of Subunit Numbers and Subunit Sizes of Enzymes: A Study of the Products of 100 Human Gene Loci', *Ann. Hum. Genet.* **39**, 383–411.

Hunt, W. G. and Selander, R. K. (1973) 'Biochemical Genetics of Hybridization in European House Mice', *Heredity* **31**, 11–33.

Hutchison, C. A. III, Newbold, J. E., Potter, S. S. and Edgell, M. H. (1974) 'Maternal Inheritance of Mammalian Mitochondrial DNA', *Nature* **251**, 536–538.

Jamieson, A. (1975) 'Enzyme Types of Atlantic Cod Stocks in the North American Banks', in *Isozymes*, Vol. IV, Ed. Markert, C. L., Academic Press, London.

Johnson, G. B. (1975) 'Use of Internal Standards in Electrophoretic Surveys of Enzyme Polymorphism', *Biochem. Genet.* **13**, 833–847.

Johnson, G. B. (1977) 'Assessing Electrophoretic Similarity: The Problem of Hidden Heterogeneity', *Ann. Rev. Ecol. Syst.* **8**, 309–328.

Johnson, M. S. (1974) 'Comparative Geographic Variation in *Menidia*', *Evolution* **28**, 607–618.

Jope, E. M. (1976) 'The Evolution of Plants and Animals under Domestication: The Contribution of Studies at the Molecular Level', *Phil. Trans. R. Soc. Lond. B.* **275**, 99–116.

Jope, M. (1969) 'The Protein of Brachiopod Shell–IV. Shell Protein from Fossil Inarticulates: Amino Acid Composition and Disc Electrophoresis of Fossil Articulate Shell Protein', *Comp Biochem. Physiol.* **30**, 225–232.

Kersters, K. and De Ley, J. (1975) 'Identification and Grouping of Bacteria by Numerical Analysis of their Electrophoretic Protein Patterns', *J. gen. Microbiol.* **87**, 333–342.

Kilgour, V. and Godfrey, D. G. (1973) 'Species-characteristic Isoenzymes of Two Aminotransferases in Trypanosomes', *Nature* **244**, 69–70.

Kilgour, V. and Godfrey, D. G. (1977) 'The Persistence in the Field of Two Characteristic Isoenzyme Patterns in Nigerian *Trypanosoma vivax*', *Ann. trop. Med. Parasit.* **71**, 387–389.

Kimura, M. (1968) 'Evolutionary Rate at the Molecular Level', *Nature* **217**, 624–626.

Kimura, M. (1977) 'Preponderance of Synonymous Changes as Evidence for the Neutral Theory of Molecular Evolution', *Nature* **267**, 275–276.

Kimura, M. and Ohta, T. (1971) 'Protein Polymorphism as a Phase of Molecular Evolution', *Nature* **229**, 467–469.

King, J. L. and Jukes, T. H. (1969) 'Non-Darwinian Evolution', *Science* **164**, 788–798.

King, J. L. and Ohta, T. H. (1975) 'Polyallelic Mutational Equilibria', *Genetics* **79**, 681–691.

King, M-C. and Wilson, A. C. (1975) 'Evolution at Two Levels in Humans and Chimpanzees', *Science* **188**, 107–116.

Kirk, R. L. (1975) 'Isozyme Variants as Markers of Population Movement in Man', in *Isozymes* Vol. IV, Ed. Markert, C. L., Academic Press, London.

Koehn, R. K. (1970) 'Functional and Evolutionary Dynamics of Polymorphic Esterases in Catostomid Fishes', *Trans. Amer. Fish. Soc.* **1**, 219–228.

Koehn, R. K. and Mitton, J. G. (1972) 'Population Genetics of Marine Pelecypods I. Ecological Heterogeneity and Evolutionary Strategy at an Enzyme Locus', *Amer. Natur.* **106**, 47–56.

Kohne, D. E., Chiscon, J. A. and Hoyer, B. H. (1972) 'Evolution of Primate DNA Sequences', *J. Hum. Evol.* **1**, 627–644.

Kung, S. D., Sakano, K., Gray, J. C. and Wildman, S. G. (1975) 'The Evolution of Fraction 1 Protein During the Origin of a New Species of *Nicotiana*', *J. Mol. Evol.* **7**, 59–64.

Lakovaara, S., Saura, A., Lankinen, P., Pohjola, L. and Lokki, J. (1976) 'The Use of Isoenzymes in Tracing Evolution and in Classifying Drosophilidae', *Zool. Scr.* **5**, 173–179.

Levin, D. A. (1977) 'The Organization of Genetic Variability in *Phlox drummondii*', *Evolution* **31**, 477–494.

Levin, D. A. and Crepet, W. L. (1973) 'Genetic Variation in *Lycopodium lucidulum:* A Phylogenetic Relic', *Evolution* **27**, 622–632.

Levinton, J. (1973) 'Genetic Variation in a Gradient of Environmental Variability: Marine Bivalvia (Mollusca)', *Science* **180**, 75–76.

Lewontin, R. C. (1974) *The Genetic Basis of Evolutionary Change*, Columbia University Press, London.

Lewontin, R. C. and Birch, L. C. (1966) 'Hybridization as a Source of Variation for Adaptation to New Environments', *Evolution* **20**, 315–336.

Longwell, A. C. (1976) 'Review of Genetic and Related Studies on Commercial Oysters and other Pelecypod Molluscs', *J. Fish. Res. Board Can.* **33**, 1100–1107.

Lucotte, G. (1977) *Le Polymorphisme Biochimique et les Facteurs de son Maintien*, Masson, Paris.

Lucotte, G. and Kaminski, M. (1976) 'Molecular Heterosis at the Conalbumin Locus in the Ring-necked Pheasant (*Phasianus colchicus*)', *Theor. Appl. Genet.* **48**, 251–253.

Lucotte, G., Kaminski, M. and Perramon, A. (1978) 'Heterosis in the Codominance Model: Electrophoretic Studies of Proteins in the Chicken–Quail Hybrid', *Comp. Biochem. Physiol.* **60B**, 169–171.

Lyddiatt, A., Peacock, D. and Boulter, D. (1978) 'Evolutionary Change in Invertebrate Cytochrome c', *J. Mol. Evol.* **11**, 35–45.

MacIntyre, R. J. (1976) 'Evolution and Ecological Value of Duplicate Genes', *Ann. Rev. Ecol. Syst.* **7**, 421–468.

Manwell, C. and Baker, C. M. A. (1963) 'A Sibling Species of Sea Cucumber Discovered by Starch Gel Electrophoresis', *Comp. Biochem. Physiol.* **10**, 39–53.

Manwell, C. and Baker, C. M. A. (1970) *Molecular Biology and the Origin of Species*, Sidgwick & Jackson, London.

Manwell, C. and Baker, C. M. A. (1975) 'Molecular Genetics of Avian Proteins XIII.

Protein Polymorphism in Three Species of Australian Passerines', *Austr. J. Biol. Sci.* **28**, 545–557.

Manwell, C. and Baker, C. M. A. (1976) 'Protein Polymorphisms in Domesticated Species: Evidence for Hybrid Origin?', in *Population Genetics and Ecology*, Ed. Karlin, S. and Nevo, E., Academic Press, London.

Marcus, N. H. (1977) 'Genetic Variation Within and Between Geographically Separate Populations of the Sea Urchin *Arbacia punctulata*', *Biol. Bull.* **153**, 560–576.

Marinković, D., Ayala, F. J. and Andjelković, M. (1978) 'Genetic Polymorphism and Phylogeny of *Drosophila subobscura*', *Evolution* **32**, 164–173.

Markert, C. L. and Møller, F. (1959) 'Multiple Forms of Enzymes: Tissue, Ontogenetic, and Species Specific Patterns', *Proc. natn. Acad. Sci. U.S.A.* **45**, 753–763.

Massaro, E. J. and Booke, H. E. (1971) 'Photoregulation of the Expression of Lactate Dehydrogenase Isozymes in *Fundulus heteroclitus* (Linnaeus)', *Comp. Biochem. Physiol.* **38B**, 327–332.

Mattei, D. M., Goldenberg, S., Morel, C., Azevedo, H. P. and Roitman, I. (1977) 'Biochemical Strain Characterization of *Trypanosoma cruzi* by Restriction Endonuclease Cleavage of Kinetoplast DNA', *FEBS Letters*, **74**, 264–268.

Maxam, A. M. and Gilbert, W. (1977) 'A New Method for Sequencing DNA', *Proc. natn. Acad. Sci. U.S.A.* **74**, 560–564.

Maxson, L. R. (1977) 'Immunological Detection of Convergent Evolution in the Frog *Anotheca spinosa*', *Syst. Zool.* **26**, 72–76.

Maxson, L. R. and Wilson, A. C. (1975) 'Albumin Evolution and Organismal Evolution in Tree Frogs (Hylidae)', *Syst. Zool.* **24**, 1–15.

Maxson, L. R., Sarich, V. M. and Wilson, A. C. (1975) 'Continental Drift and the Use of Albumin as an Evolutionary Clock', *Nature* **255**, 397–399.

Mayr, E. (1969a) 'The Role of Systematics in Biology' in *Systematic Biology*, National Academy of Sciences, Washington.

Mayr, E. (1969b) *Principles of Systematic Zoology*, McGraw-Hill, London.

Mayr, E. (1974) 'Cladistic Analysis or Cladistic Classification', *Z. zool. Syst. Evolut.-forsch.* **12**, 94–128.

Merritt, R. B. (1972) 'Geographic Distribution and Enzymatic Properties of Lactate Dehydrogenase Allozymes in the Fathead Minnow, *Pimephales promelas*', *Amer. Natur.* **106**, 173–184.

Merritt, R. B., Rogers, J. F. and Kurz, B. J. (1978) 'Genic Variability in the Longnose Dace, *Rhinichthys cataractae*', *Evolution* **32**, 116–124.

Mickevich, M. F. and Johnson, M. S. (1976) 'Congruence between Morphological and Allozyme Data in Evolutionary Inference and Character Evolution', *Syst. Zool.* **25**, 260–270.

Miles, M. A., Souza, D., Povoa, M., Shaw, J. L., Lainson, R. and Toye, P. J. (1978) 'Isozymic Heterogeneity of *Trypanosoma cruzi* in the First Autochthonous Patients with Chagas' Disease in Amazonian Brazil', *Nature* **272**, 819–821.

Mitchell, W. M. (1967) 'A Potential Source of Electrophoretic Artifacts in Polyacrylamide Gels', *Biochem. Biophys. Acta.* **147**, 171–174.

Mitton, J. B. (1978) 'Relationship between Heterozygosity for Enzyme Loci and Variation of Morphological Characters in Natural Populations', *Nature* **273**, 661–662.

Moav, R., Brody, T., Wohlfarth, G. and Hulata, G. (1976) 'Application of Electrophoretic Genetic Markers to Fish Breeding. I. Advantages and Methods', *Aquaculture* **9**, 217–228.

Moore, G. W. (1976) 'Proof for the Maximum Parsimony ('Red King') Algorithm', in *Molecular Anthropology*, Ed. Goodman, M. and Tashian, R. E., Plenum Press, London.

Moore, G. W., Goodman, M., Callahan, C., Holmquist, R. and Moise, H. (1976) 'Stochastic versus Augmented Maximum Parsimony Method for Estimating Superimposed Mutations in the Divergent Evolution of Protein Sequences', *J. Mol. Biol.* **105**, 15–38.

Mross, G. A. and Doolittle, R. F. (1967) 'Amino Acid Sequence Studies on Artiodactyl Fibrinopeptides. II. Vicuna, Elk, Muntjak, Pronghorn Antelope and Water Buffalo. *Archs. Biochem. Biophys.* **122**, 674–684.

Murdock, E. A., Ferguson, A. and Seed, R. (1975) 'Geographical Variation in Leucine Aminopeptidase in *Mytilus edulis* L. from the Irish Coasts', *J. exp. mar. Biol. Ecol.* **19**, 33–41.

Murray, R. A. and Solomon, M. G. (1978) 'A Rapid Technique for Analysing Diets of Invertebrate Predators by Electrophoresis', *Ann. appl. Biol.* **90**, 7–10.

Nei, M. (1972) 'Genetic Distance between Populations', *Amer. Natur.* **106**, 283–292.

Nei, M. (1975) *Molecular Population Genetics and Evolution*, North-Holland, Amsterdam.

Nei, M. (1976) in *Population Genetics and Ecology*, Ed. Karlin, S. and Nevo, E., Academic Press, London.

Nei, M. and Roychoudhury, A. K. (1974) 'Genic Variation Within and Between the Three Major Races of Man, Caucasoids, Negroids, and Mongoloids', *Amer. J. Hum. Genet.* **26**, 421–443.

Nevo, E. (1976) 'Adaptive Strategies of Genetic Systems in Constant and Varying Environments', in *Population Genetics and Ecology*, Ed. Karlin, S. and Nevo, E., Academic Press, London.

Nevo, E. and Barr, Z. (1976) 'Natural Selection of Genetic Polymorphisms along Climatic Gradients', in *Population Genetics and Ecology*, Academic Press, London.

Nevo, E. and Shaw, C. R. (1972) 'Genetic Variation in a Subterranean Mammal, *Spalax ehrenbergi*', *Biochem. Genet.* **7**, 235–241.

Nolan, R. A., Brush, A. H., Arnheim, N. and Wilson, A. C. (1975) 'An Inconsistency between Protein Resemblance and Taxonomic Resemblance: Immunological Comparison of Diverse Proteins from Gallinaceous Birds', *Condor* **77**, 154–159.

Northcote, T. G., Williscroft, S. N. and Tsuyuki, H. (1970) 'Meristic and Lactate Dehydrogenase Genotype Differences in Stream Populations of Rainbow Trout below and above a Waterfall', *J. Fish. Res. Board. Can.* **27**, 1987–1995.

Nuttall, G. H. F. (1901) 'The New Biological Test for Blood in Relation to Zoological Classification', *Proc. Roy. Soc. Lond.* **69**, 150–153.

Nygren, A., Nyman, L., Svensson, K. and Jahnke, G. (1975) 'Cytological and Biochemical Studies in Back-crosses between the Hybrid Atlantic Salmon × Sea Trout and its Parental Species', *Hereditas* **81**, 55–62.

Nyman, L. (1970) 'Electrophoretic Analysis of Hybrids between Salmon (*Salmo salar* L.) and Trout (*Salmo trutta* L.). *Trans. Am. Fish. Soc.* **99**, 229–236.

Oelshlegel, F. J. and Stahmann, M. A. (1973) 'The Electrophoretic Technique–a Practical Guide for its Application', *Bull. Torrey Bot. Club.* **100**, 260–271.

O'Farrell, P. H. (1975) 'High Resolution Two-dimensional Electrophoresis of Proteins', *J. Biol. Chem.* **250**, 4007–4021.

Ogihara, M. (1975) 'Correlation between Lactate Dehydrogenase Isozyme Patterns of Mammalian Livers and Dietary Environments', in *Isozymes*, Vol. IV, Ed. Markert, C. L., Academic Press, London.

Ohno, S. (1970) *Evolution by Gene Duplication*, Allen & Unwin, London.

Ohta, T. and Kimura, M. (1971) 'Behaviour of Neutral Mutants Influenced by Associated Over-dominant Loci in Finite Populations', *Genetics* **69**, 247–260.

Parker, E. D. and Selander, R. K. (1976) 'The Organization of Genetic Diversity in the Parthenogenetic Lizard *Cnemidophorus tesselatus*', *Genetics*, **84**, 791–805.

Patton, J. L. and Yang, S. Y. (1977) 'Genetic Variation in *Thomomys bottae* Pocket Gophers: Macrogeographic Patterns', *Evolution* **31**, 697–720.

Patton, J. L., MacArthur, H. and Yang, S. Y. (1976) 'Systematic Relationships of the Four-Toed Populations of *Dipodomys heermanni*', *J. Mammal.* **57**, 159–163.

Payne, R. H., Child, A. R. and Forrest, A. (1971) 'Geographic Variation in the Atlantic Salmon', *Nature* **231**, 250–252.

Payne, R. H., Child, A. R. and Forrest, A. (1972) 'The Existence of Natural Hybrids between the European Trout and the Atlantic Salmon', *J. Fish Biol.* **4**, 233–236.

Potter, S. S., Newbold, J. F., Hutchison, C. A. III and Edgell, M. H. (1975) 'Specific Cleavage Analysis of Mammalian Mitochondrial DNA', *Proc. natn. Acad. Sci. U.S.A.* **72**, 4496–4500.

Powell, J. R. (1975) 'Isozymes and Non-Darwinian Evolution: A Re-Evaluation', in *Isozymes*, Vol. IV, Ed. Markert, C. L., Academic Press, London.

Power, H. W. (1975) 'A Model of how the Sickle-cell Gene produces Malaria Resistance', *J. theor. Biol.* **50**, 121–127.

Powers, D. A. and Powers, D. (1975) 'Predicting Gene Frequencies in Natural Populations: A Testable Hypothesis', in *Isozymes*, Vol. IV, Ed. Markert, C. L., Academic Press, London.

Prager, E. M. and Wilson, A. C. (1978) 'Construction of Phylogenetic Trees for Proteins and Nucleic Acids: Empirical Evaluation of Alternative Matrix Methods', *J. Mol. Evol.* **11**, 129–142.

Prager, E. M., Welling, G. W. and Wilson, A. C. (1978) 'Comparison of Various Immunological Methods for Distinguishing among Mammalian Pancreatic Ribonucleases of Known Amino Acid Sequence', *J. Mol. Evol.* **10**, 293–308.

Reinitz, G. L. (1977) 'Tests for Association of Transferrin and Lactate Dehydrogenase Phenotypes with Weight Gain in Rainbow Trout (*Salmo gairdneri*)', *J. Fish. Res. Board. Can.* **34**, 2333–2337.

Richardson, R. H. and Smouse, P. E. (1976) 'Patterns of Molecular Variation. I. Interspecific Comparisons of Electromorphs in the *Drosophila mulleri* Complex', *Biochem. Genet.* **14**, 447–466.

Rogers, J. S. (1972) 'Measures of Genetic Similarity and Genetic Distance', *Univ. Texas Stud. Genet.* **VII**, 145–153.

Romero-Herrera, A. E., Lehmann, H., Joysey, K. A. and Friday, A. E. (1978) 'On the Evolution of Myoglobin', *Phil. Trans. Royal Soc. Lond.B* **283**, 61–163.

Ross, G. C. (1976) 'Isoenzymes in *Schistosoma* spp.: LDH, MDH and Acid Phosphatases Separated by Isoelectric Focusing in Polyacrylamide Gel', *Comp. Biochem. Physiol.* **55B**, 343–346.

Sanger, F. and Coulson, A. R. (1975) 'A Rapid Method for Determining Sequences in DNA by Primed Synthesis with DNA Polymerase', *J. molec. Biol.* **94**, 441–448.

Sargent, J. R. and George, S. G. (1975) *Methods in Zone Electrophoresis*, B. D. H. Chemicals, Poole.

Sarich, V. M. (1973) 'The Giant Panda is a Bear', *Nature* **245**, 218–220.

Sarich, V. M. (1977) 'Rates, Sample Sizes, and the Neutrality Hypothesis for Electrophoresis in Evolutionary Studies', *Nature* **265**, 24–28.

Schnell, G. D., Best, T. L. and Kennedy, M. L. (1978) 'Interspecific Morphologic Variation in Kangaroo Rats (*Dipodomys*): Degree of Concordance with Genic Variation', *Syst. Zool.* **27**, 34–48.

Scopes, R. K. (1968) 'Methods for Starch Gel Electrophoresis of Scarcoplasmic Proteins', *Biochem. J.* **107**, 139–150.

Selander, R. K. (1976) 'Genic Variation in Natural Populations', in *Molecular Evolution*, Ed. Ayala, F. J., Sinauer Associates, Sunderland, Massachusetts.

Selander, R. K. and Kaufman, D. W. (1975) 'Genetic Population Structure and Breeding Systems', in *Isozymes*, Vol. IV, Ed. Markert, C. L., Academic Press, London.

Selander, R. K., Hunt, W. G. and Yang, S. Y. (1969) 'Protein Polymorphism and Genic Heterozygosity in Two European Subspecies of the House Mouse', *Evolution* **23**, 379–390.

Selander, R. K., Yang, S. Y., Lewontin, R. C. and Johnson, W. E. (1970) 'Genetic Variation in the Horseshoe Crab (*Limulus polyphemus*), a phylogenetic 'relic'', *Evolution* **24**, 402–414..

Serov, O. L., Zakijan, S. M. and Kulichov, V. A. (1978) 'Allelic Expression in Intergeneric Fox Hybrids (*Alopex lagopus* × *Vulpes vulpes*). III. Regulation of the Expression of the Parental Alleles at the *Gpd* Locus Linked to the X Chromosome', *Biochem. Genet.* **16**, 145–157.

Shaklee, J. B. (1975) 'The Role of Subunit Interactions in the Genesis of Non-Binomial Lactate Dehydrogenase Isozyme Distributions', in *Isozymes*, Vol. I, Ed. Markert, C. L., Academic Press, London.

Shaklee, J. B., Christiansen, J. A., Sidell, B. D., Prosser, C. L. and Whitt, G. S. (1977) 'Molecular Aspects of Temperature Acclimination in Fish: Contributions of Changes in Enzyme Activities and Isozyme Patterns to Metabolic Reorganization in the Green Sunfish', *J. exp. Zool.* **201**, 1–20.

Shaw, C. R. and Koen, A. L. (1963) 'Hormone-Induced Esterase in Mouse Kidney', *Science* **140**, 70–71.

Shick, J. M. and Lamb, A. N. (1977) 'Asexual Reproduction and Genetic Population Structure in the Colonizing Sea Anemone *Haliplanella luciae*', *Biol. Bull.* **153**, 604–617.

Shields, G. F. and Straus, N. A. (1975) 'DNA-DNA Hybridization Studies of Birds', *Evolution* **29**, 159–166.

Short, L. L. (1973) 'Hybridization, Taxonomy and Avian Evolution', *Ann. Missouri Bot. Gard.* **59**, 447–453.

Sibley, C. G. (1954) 'Hybridization in the Red-eyed Towhees of Mexico', *Evolution* **8**, 252–290.

Sibley, C. G. (1960) 'The Electrophoretic Patterns of Avian Egg-white Proteins as Taxonomic Characters', *Ibis* **102**, 215–284.

Sibley, C. G. (1970) 'A Comparative Study of the Egg-white Proteins of Passerine Birds', *Bull. Peabody Mus. nat. Hist.* **32**, 1–131.

Sibley, C. G. (1976) 'Protein Evidence of the Relationships of Some Australian Passerine Birds', *Proc. Int. Ornithological Congress.* **16**, 557–570.

Sibley, C. G. and Ahlquist, J. E. (1972*a*) 'A Comparative Study of the Egg-white Proteins of Non-Passerine Birds', *Bull. Peabody Mus. nat. Hist.* **39**, 1–276.

Sibley, C. G. and Brush, A. H. (1967) 'An Electrophoretic Study of Avian Eye-lens Proteins', *Auk* **84**, 203–219.

Sibley, C. G. and Frelin, C. (1972*b*) 'The Egg-white Protein Evidence for Ratite Affinities', *Ibis* **114**, 377–387.

Simoncsits, A., Brownlee, G. G., Brown, R. S., Rubin, J. R. and Guilley, H. (1977) 'New Rapid Gel Sequencing Method for RNA', *Nature* **269**, 833–836.

Simpson, G. G. (1961) *Principles of Animal Taxonomy*, Columbia University Press, New York.

Singh, R. S., Lewontin, R. C. and Felton, A. A. (1976) 'Genetic Heterogeneity within Electrophoretic 'Alleles' of Xanthine Dehydrogenase in *Drosophila pseudoobscura*', *Genetics* **81**, 609–629.

Smith, I. (1976) *Chromatographic and Electrophoretic Techniques*, Vol. II, *Zone Electrophoresis*, Heinemann, London.

Smith, M. H., Garten, C. T. and Ramsey, P. R. (1975) 'Genic Heterozygosity and Population Dynamics in Small Mammals', in *Isozymes*, Vol. IV, Ed. Markert, C. L., Academic Press, London.

Smith, P. M. (1976) *The Chemotaxonomy of Plants*, Edward Arnold, London.

Sneath, P. H. A. and Sokal, R. R. (1973) *Numerical Taxonomy*, W. H. Freeman & Co., San Francisco.

Snyder, T. P. (1977) 'A New Electrophoretic Approach to Biochemical Systematics of Bees', *Biochem. Syst. Ecol.* **5**, 133–150.

Sokal, R. R. and Rohlf, F. J. (1969) *Biometry*, W. H. Freeman & Co., San Francisco.

Solomon, D. J. and Child, A. R. (1978) 'Identification of Juvenile Natural Hybrids between Atlantic Salmon (*Salmo salar* L.) and trout (*Salmo trutta* L.)', *J. Fish Biol.* **12**, 499–501.

Sonderegger, P. and Christen, P. (1978) 'Comparison of the Evolution Rates of Cytosolic and Mitochnodrial Aspartate Aminotransferase', *Nature* **275**, 157–159.

Stansfield, W. D. (1977) *The Science of Evolution*, Collier Macmillan, London.

Steer, M. W. (1975) 'Evolution in the Genus *Avena*: Identification of Different Forms of Ribulose Diphosphate Carboxylase', *Can. J. Genet. Cytol.* **17**, 337–344.

Steer, M. W. and Kernoghan, D. (1977) 'Nuclear and Cytoplasmic Genome Relationships in the Genus *Avena:* Analysis by Isoelectric Focusing of Ribulose Biphosphate Carboxylase Subunits', *Biochem. Genet.* **15**, 273–286.

Stegeman, H. (1977) 'Plant Proteins Evaluated by Two-dimensional Methods', in *Electrofocusing and Isotachophoresis*, Ed. Radola, B. J. and Graesslin, D., Walter de Gruyter, Berlin.

Stern, C. (1970) 'Model Estimates of the Number of Gene Pairs Involved in Pigmentation Variability of the Negro-American', *Hum. Hered.* **20**, 165–168.

Stratil, A. (1967) 'The Effect of Iron Addition to Avian Egg White on the Behaviour of Conalbumin Fractions in Starch Gel Electrophoresis', *Comp. Biochem. Physiol.* **22**, 227–233.

Suomalainen, E. and Saura, A. (1973) 'Genetic Polymorphism and Evolution in Parthenogenetic Animals. I. Polyploid Curculionidae', *Genetics* **74**, 489–508.

Tait, A. (1970) 'Enzyme Variation between Syngens in *Paramecium aurelia*', *Biochem. Genet.* **4**, 461–470.

Tåning, Å. V. (1952) 'Experimental Study of Meristic Characters in Fishes', *Biol. Rev.* **27**, 169–193.

Tucker, J. M. (1952) 'Evolution of the California Oak *Quercus Alvordiana*', *Evolution* **6**, 162–180.

Turner, B. J. (1974) 'Genetic Divergence of Death Valley Pupfish Populations: Species-Specific Esterases', *Comp. Biochem. Physiol.* **46B**, 57–70.

Uzzell, T. and Berger, L. (1975) 'Electrophoretic Phenotypes of *Rana ridibunda*, *Rana lessonae*, and their hybridogenetic associate, *Rana esculenta*', *Proc. Acad. Nat. Sci. Philadelphia* **127**, 13–24.

Vessel, E. S. (1975) 'Medical Uses of Isozymes', in *Isozymes*, Vol. II, Ed. Markert, C. L., Academic Press, London.

Vrijenhoek, R. C. (1975*a*) 'Effects of Parasitism on the Esterase Isozyme Patterns of Fish Eyes', *Comp. Biochem. Physiol.* **50B**, 75–76.

Vrijenhoek, R. C. (1975*b*) 'Gene Dosage in Diploid and Triploid Unisexual Fishes (*Poeciliopsis*, Poecilidae), in *Isozymes*, Vol. IV, Ed. Markert, C. L., Academic Press, London.

Vrijenhoek, R. C., Angus, R. A. and Schultz, R. J. (1977) 'Variation and Heterozygosity in Sexually vs. Clonally Reproducing Populations of *Poeciliopsis*', *Evolution* **31**, 767–781.

Vrijenhoek, R. C., Angus, R. A. and Schultz, R. J. (1978) 'Variation and Clonal Structure in a Unisexual Fish', *Amer. Natur.* **112**, 41–55.

Wallace, D. G. and Boulter, D. (1976) 'Immunological Comparisons of Higher Plant Plastocyanins', *Phytochemistry* **15**, 137–141.

Ward, R. D. and Beardmore, J. A. (1977) 'Protein Variation in the Plaice *Pleuronectes platessa* L.', *Genet. Res. Camb.* **30**, 45–62.

White, M. J. D. (1978) *Modes of Speciation*, W. H. Freeman & Co., San Francisco.

Whitt, G. S., Childers, W. F. and Cho, P. L. (1973) 'Allelic Expression at Enzyme Loci in an Intertribal Hybrid Sunfish', *J. Hered.* **64**, 55–61.

Whitt, G. S., Philipp, D. P. and Childers, W. F. (1977) 'Aberrant Gene Expression during the Development of Hybrid Sunfishes (Perciformes, Teleostei)', *Differentiation* **9**, 97–109.

Wilkins, N. P. (1975) 'Phosphoglucose Isomerase in Marine Molluscs', in *Isozymes*, Vol. IV, Ed. Markert, C. L., Academic Press, London.

Wilkinson, J. H. (1970) *Isoenzymes*, Chapman & Hall, London.

Williams, G. C., Koehn, R. K. and Mitton, J. B. (1973) 'Genetic Differentiation without Isolation in the American Eel, *Anguilla rostrata*', *Evolution* **27**, 192–204.

Wilson, A. C. (1975) 'Relative Rates of Evolution of Organisms and Genes', *Stadler Genet. Symp.* **7**, 117–134.

Wilson, A. C. (1976) 'Gene Regulation in Evolution; in *Molecular Evolution*, Ed. Ayala, F. J., Sinauer Associates, Sunderland, Massachusetts.

Wilson, A. C., Bush, G. L., Case, S. M. and King, M.-C. (1975) 'Social Structuring of Mammalian Populations and Rate of Chromosomal Evolution', *Proc. natn. Acad. Sci. U.S.A.* **72**, 5061–5065.

Wilson, A. C., Carlson, S. S. and White, T. J. (1977) 'Biochemical Evolution', *Ann. Rev. Biochem.* **46**, 573–639.

Wilson, A. C., Maxson, L. R. and Sarich, V. M. (1974) 'Two Types of Molecular Evolution. Evidence from Studies of Interspecific Hybridization', *Proc. natn. Acad. Sci. U.S.A.* **71**, 2843–2847.

Wilson, E. O., Eisner, T., Briggs, W. R., Dickerson, R. E., Metzenberg, R. L., O'Brien, R. D., Susman, M. and Bogg, W. E. (1973) *Life on Earth*, Sinauer Associates, Sunderland, Connecticut.

Workman, P. L. and Niswander, J. D. (1970) 'Populations Studies on Southwestern Indian Tribes. II. Local Genetic Differentiation in the Papago', *Amer. J. Hum. Genet.* **22**, 24–49.

Wright, C. A. (1974) *Biochemical and Immunological Taxonomy of Animals*, Academic Press, London.

Index